FUNDAMENTALS OF
MINING GEOLOGY

FUNDAMENTALS OF
MINING GEOLOGY

Ravindra Singh

Oxford Book Company
Jaipur ▪ New Delhi

Published By

Oxford Book Company

Regd. Off. "Bony Residency", Gate No. 2
Opp. Tilak Public School, Vishweswariya Nagar,
Gopalpura Road, Jaipur - 302018, India
Ph: 0141 - 2761280, Telefax: 2761381
email - oxfordbookjpr@gmail.com
website - oxfordbookcompany.com

Delhi Office: 102, 1st Floor, Satyam House,
4327/3, Ansari Road, Darya Ganj,
New Delhi - 110002, India
Ph: 011-45652440

Distributed by
KEYA INTERNATIONAL PRESS LLP
E-mail: keyainternationalpressllp@gmail.com

Published in India, 2026

ISBN : 9789355240019

Typeset by Shivangi Computers, Jaipur

Printed and bound in India by Replika Press Pvt. Ltd.

Contents

Preface

Mineralogy is the study of chemistry, crystal structure, and physical properties of minerals. Specific studies within mineralogy include the processes of mineral origin and formation, classification of minerals, their geographical distribution, as well as their utilization. Early writing on mineralogy, especially on gemstones, comes from ancient Babylonia, the ancient Greco-Roman world, ancient and medieval China, and Sanskrit texts from ancient India. Books on the subject included the Naturalis Historia of Pliny the Elder which not only described many different minerals but also explained many of their properties. The German Renaissance specialist Georgius Agricola wrote works such as *De re metallica* and *De Natura Fossilium* which begin the scientific approach to the subject. Systematic scientific studies of minerals and rocks developed in post-Renaissance Europe. The modern study of mineralogy was founded on the principles of crystallography and microscopic study of rock sections with the invention of the microscope in the 17th century.

In mines which use rubber tired equipment for coarse ore removal, the ore is removed from the stope using centre articulated vehicles. These pieces of equipment may operate using diesel or electric engines and resemble a low-profile front end loader. The ore is then dumped into a truck to be hauled to the surface. In deeper mines the ore is dumped down an ore pass where it falls to a collection level. On the collection level, it may receive primary crushing via jaw or cone crusher. The ore is then moved by conveyor belts, trucks or occasionally trains to the shaft to be hoisted to the surface in buckets or skips and emptied into bins beneath the surface headframe for transport to the mill.

In some cases the underground primary crusher feeds an inclined conveyor belt which delivers ore via an incline shaft direct to the surface. The ore is fed down ore passes, with mining equipment accessing the ore body via a decline from surface.

Underground mining refers to a group of underground mining techniques used to extract coal, oil shale and other minerals or geological materials from sedimentary rocks. Because deposits in sedimentary rocks are commonly layered and relatively less hard, the mining methods used differ from those used to mine deposits in igneous or metamorphic rocks. Underground mining techniques also differ greatly from those of surface mining.

The book incorporates salient features of the earth—its structure and composition.

— ***Ravindra Singh***

Chapter 1

Mining Techniques

Open-pit Mining

Open-pit mining or opencast mining refers to a method of extracting rock or minerals from the earth by their removal from an open pit or borrow.

Figure 1 : El Chino, located near Silver City, New Mexico, is an open-pit copper mine.

The term is used to differentiate this form of mining from extractive methods that require tunnelling into the earth. Open-pit mines are used when deposits of commercially useful minerals or rock are found near the surface; that is, where the overburden (surface material covering the valuable deposit) is relatively thin or the material of interest is structurally unsuitable for tunnelling (as would be the case for sand, cinder, and gravel). For minerals that occur deep below the surface—where the overburden is thick or the mineral occurs as veins in hard rock—underground mining methods extract the valued material.

Open-pit mines that produce building materials and dimension stone are commonly referred to as quarries. People are unlikely to make a distinction between an open-pit mine and other types of open-cast mines, such as quarries, borrows, placers, and strip mines.

Open-pit mines are typically enlarged until either the mineral resource is exhausted, or an increasing ratio of overburden to ore makes further mining uneconomic. When this occurs, the exhausted mines are sometimes converted to landfills for disposal of solid wastes. However, some form of water control is usually required to keep the mine pit from becoming a lake.

Extraction

Open-pit mines are dug on benches, which describe vertical levels of the hole. These benches are usually on four metre to sixty metre intervals, depending on the size of the machinery that is being used. Many quarries do not use benches, as they are usually shallow.

Most walls of the pit are generally dug on an angle less than vertical, to prevent and minimise damage and danger from rock falls. This depends on how weathered the rocks are, and the type of rock, and also how many structural weaknesses occur within the rocks, such as a fault, shears, joints or foliations.

The walls are stepped. The inclined section of the wall is known as the batter, and the flat part of the step is known as the bench or berm. The steps in the walls help prevent rock falls continuing down the entire face of the wall. In some instances additional ground support is required and rock bolts, cable bolts and shotcrete are used. De-watering bores may be used to relieve water pressure by drilling horizontally into the wall, which is often enough to cause failures in the wall by itself.

A haul road is usually situated at the side of the pit, forming a ramp up which trucks can drive, carrying ore and waste rock.

Waste rock is piled up at the surface, near the edge of the open pit. This is known as the waste dump. The waste dump is also tiered and stepped, to minimise degradation. Ore which has been processed is known as tailings, and is generally a slurry. This is pumped to a tailings dam or settling pond, where the water evaporates. Tailings dams can often be toxic due to the presence of unextracted sulfide minerals, some forms of toxic minerals in the gangue, and often cyanide which is used to treat gold ore via the cyanide leach process. This toxicity has the potential to negatively impact on the surrounding environment.

Surface Mining

Surface mining (also commonly called strip mining, though this is actually only one possible form of surface mining), is a type of mining in which soil and rock overlying the mineral deposit (the overburden) are removed. It is the opposite of underground mining, in which the overlying rock is left in place, and the mineral removed through shafts or tunnels.

Surface mining began in the mid-sixteenth century and is practiced throughout the world, although the majority of surface mining occurs in North America. It gained popularity throughout the 20th century, and is now the predominant form of mining in coal beds such as those in Appalachia and America's Midwest.

In most forms of surface mining, heavy equipment, such as earthmovers, first remove the overburden. Next, huge machines, such as dragline excavators or Bucket wheel excavators, extract the mineral.

Strip Mining

"Strip mining" is the practice of mining a seam of mineral by first removing a long strip of overlying soil and rock (the overburden). It is most commonly used to mine coal or tar sand. Strip mining is only practical when the ore body to be excavated is relatively near the surface. This type of mining uses some of the largest machines on earth, including bucket-wheel excavators which can move as much as 12,000 cubic meters of earth per hour.

There are two forms of strip mining. The more common method is "area stripping", which is used on fairly flat terrain, to extract deposits over a large area. As each long strip is excavated, the overburden is placed in the excavation produced by the previous strip.

"Contour stripping" involves removing the overburden above the mineral seam near the outcrop in hilly terrain, where the mineral outcrop usually follows the contour of the land. Contour stripping is often followed by auger mining into the hillside, to remove more of the mineral. This method commonly leaves behind terraces in mountainsides.

Among others, strip mining is used to extract the oil-impregnated sand in the Athabasca Tar Sands in Alberta. It is also common in coal mining. Bucket-wheel excavators are widely used for this purpose, however, they are prone to damage and require many millions of dollars to repair.

Open-pit Mining

"Open-pit mining" refers to a method of extracting rock or minerals from the earth through their removal from an open pit or borrow. Although open-pit mining is sometimes mistakenly referred to as "strip mining", the two methods are different.

Mountaintop Removal

"Mountaintop removal mining" (MTR) is a form of coal mining that uses explosives to blast "overburden" off the top of some Appalachian mountains. Excess mining waste or "overburden" is dumped by large trucks into fills in nearby holler or valley fills. MTR involves the mass restructuring of earth in order to reach the coal seam as deep as 400 feet (120 m) below the surface.

Mountaintop removal replaces previously steep forested topography with government approved post mining reclamation land uses. Economic development attempts on reclaimed mine sites include prisons such the Big Sandy Federal Penitentiary in Martin County, Kentucky, small town airports, golf courses such as Twisted Gun in Mingo County, West Virginia and Stonecrest Golf Course in Floyd County, Kentucky, as well as industrial scrubber sludge disposal sites, solid waste landfills, trailer parks, explosive manufacturers, and storage rental lockers.

The technique has been used increasingly in recent years in the Appalachian coal fields of West Virginia, Kentucky, Virginia and Tennessee in the United States. The profound changes in topography and disturbance of pre-existing ecosystems have made mountaintop removal highly controversial.

Advocates of mountaintop removal point out that once the areas are reclaimed as mandated by law, the technique provides premium flat land suitable for many uses in a region where flat land is at a premium. They also maintain that the new growth on reclaimed mountaintop mined areas is better able to support populations of game animals.

Critics contend that mountaintop removal is a disastrous practice that benefits a small number of corporations at the expense of local communities and the environment. A U.S. Environmental Protection Agency (EPA) environmental impact statement finds that streams near valley fills sometimes may contain higher levels of minerals in the water and decreased aquatic biodiversity. The statement also estimates that 724 miles (1,165 km) of Appalachian streams were buried by valley fills from 1985 to 2001.

Blasting at a mountaintop removal mine expels dust and fly-rock into the air, which can then disturb or settle onto private property nearby. This dust may contains sulfur compounds, which some claim corrode structures and tombstones and is a health hazard.

Although MTR sites are required to be reclaimed after mining is complete, reclamation has traditionally focused on stabilizing rock and controlling erosion, but not always on reforesting the area. Quick-growing, non-native grasses, planted to quickly provide vegetation on a site, compete with tree seedlings, and trees have difficulty establishing root systems in compacted backfill. Consequently, biodiversity suffers in a region of the United States with numerous endemic species. Erosion also increases, which can intensify flooding. In the Eastern United States, the Appalachian Regional Reforestation Initiative works to promote the use of trees in mining reclamation.

Dredging

"Dredging" is a method often used to bring up underwater mineral deposits. Although dredging is usually employed to clear or enlarge waterways for boats, it can also recover significant amounts of underwater minerals relatively efficiently and cheaply.

Highwall Mining

Highwall mining is another form of surface mining that evolved from auger mining. In Highwall mining, the coal seam is penetrated by a continuous miner propelled by a hydraulic Pushbeam Transfer Mechanism (PTM). A typical cycle includes sumping (launch-pushing forward) and shearing (raising and lowering the cutterhead boom to cut the entire height of the coal seam). As the coal recovery cycle continues, the cutterhead is progressively launched into the coal seam for 19.72 feet (6.01 m). Then, the Pushbeam Transfer Mechanism

(PTM) automatically inserts a 19.72-foot (6.01 m) long rectangular Pushbeam (Screw-Conveyor Segment) into the centre section of the machine between the Powerhead and the cutterhead.

The Pushbeam system can penetrate nearly 1,000 feet (300 m) into the coal seam. One patented Highwall mining systems use augers enclosed inside the Pushbeam that prevent the mined coal from being contaminated by rock debris during the conveyance process.

Using a video imaging and/or a gamma ray sensor and/or other Geo-Radar systems like a coal-rock interface detection sensor (CID), the operator can see ahead projection of the seam-rock interface and guide the continuous miner's progress. Highwall mining can produce thousands of tons of coal in contour-strip operations with narrow benches, previously mined areas, trench mine applications and steep-dip seams with controlled water-inflow pump system and/or a gas (inert) venting system.

Recovery is much better than Augering, but the mapping of areas that have been developed by a Highwall miner are not mapped as rigorously as deep mined areas. Very little soil is displaced in contrast with mountain top removal; however a large amount of capital is required to operate and own a Highwall miner. But then this Highwall mining system is the innovative roadmap future potential and stay or being better competitive in the area of environmental friendly non mountain-top (overburden) removal operated by only 4 crew members.

Mapping of the outcrop as well as core hole data and samples taken during the bench making process are taken into account to best project the panels that the Highwall miner will cut. Obstacles that could be potentially damaged by subsidence and the natural contour of the Highwall are taken into account, and a surveyor points the Highwall miner in a line (Theoretical Survey Plot-Line) mostly perpendicular to the Highwall.

Parallel lines represent the drive cut into the mountain (up to 1,000 feet (300 m) deep), without heading or corrective steering actuation on a navigation Azimuth during mining results in missing a portion of the coal seam and is a potential danger of cutting in pillars from previous mined drives due to horizontal drift (Roll) of the Pushbeam-Cuttermodule string. Recently Highwall miners have penetrated more than 1050 feet into the coal seam, and today's models are capable of going farther, with the support of gyro navigation and not limited anymore by the amount of cable stored on the machine. The maximum depth would be determined by the stress of further penetration and associated power draw, but today's optimized Pushbeam Discrete Element Modelling (DEM) shows smart-drive extended penetrations are possible.

Environmental and Health Issues

The large impact of surface mining on the topography, vegetation, and water resources has made it highly controversial.

Surface mining is subject to state and federal reclamation requirements, but adequacy of the requirements is a constant source of contention. Unless reclaimed, surface mining can leave behind large areas of infertile waste rock, as 70% of material excavated is waste.

In the United States, the Surface Mining Control and Reclamation Act of 1977 mandates reclamation of surface coal mines. Reclamation for non-coal mines is regulated by state and local laws, which may vary widely.

Human Health

The United Mine Workers of America has spoken against the use of human sewage sludge to reclaim surface mining sites in Appalachia. The UMWA launched its campaign against the use of sludge on mine sites in 1999 after eight UMWA workers became ill from exposure to Class B sludge spread near their workplace.

On August 20, 2004 at 2:30 a.m. a boulder accidentally pushed off an A&G Coal surface mine above the town of Inman, Virginia rolled 649 feet (198 m) down the mountain and into a home. Three-year-old Jeremy Davidson was crushed in his bed while he slept. The Davidson family settled with A&G Coal for $3 million in 2006, and left the region.

Environmental Impact

According to a 2010 report in the journal *Science*, mountaintop mining has caused numerous environmental problems which mitigation practices have not successfully addressed. For example, valley fills frequently bury headwater streams causing permanent loss of ecosystems. In addition, the destruction of large tracts of deciduous forests has threatened several endangered species and led to a loss of biodiversity.

Placer Mining

Placer mining is the mining of alluvial deposits for minerals. This may be done by open-pit (also called open-cast mining) or by various surface excavating equipment or tunnelling equipment.

The name derives from Spanish, *placer*, meaning "sandbank." It refers to mining the precious metal deposits (particularly gold and gemstones) found in alluvial deposits—deposits of sand and gravel in modern or ancient stream beds.

The metal or gemstones, having been moved by stream flow from an original source such as a vein, is typically only a minuscule portion of the total deposit. Since gems and heavy metals like gold are considerably more dense than sand, they tend to accumulate at the base of placer deposits.

The containing material may be too loose to safely mine by tunnelling. Where water under pressure is available, it may be used to mine, move, and separate the precious material from the deposit, a method known as hydraulic mining, hydraulic sluicing or hydraulicking.

History

Placers supplied most of the gold for a large part of the ancient world. Hydraulic mining methods such as hushing were used widely by the Romans across their empire, but especially in the gold fields of northern Spain after its conquest by Augustus in 25 BC. One of the largest sites was at Las Médulas, where seven 30 mile long aqueducts were used to work the alluvial gold deposits through the first century AD. (Inclusions of platinum-group metals in a very large proportion of gold items indicate that the gold was largely derived from placer or alluvial deposits.

Platinum group metals are seldom found with gold in hardrock reef or vein deposits.) In North America, placer mining was famous in the context of several gold rushes, particularly the California Gold Rush, the Fraser Canyon Gold Rush and the Klondike Gold Rush.

Placer mining continues in many areas of the world as a source of diamonds, industrial minerals and metals, gems (in Myanmar and Sri Lanka), platinum, and of gold (in the Yukon, Alaska and British Columbia).

Methods

A number of methods are used to mine placer gold.

Panning

The simplest technique to extract gold from placer ore is panning. In panning, some mined ore is placed in a large metal or plastic pan, combined with a generous amount of water, and agitated so that the gold particles, being of higher density than the other material, settle to the bottom of the pan. The lighter gangue material such as sand, mud and gravel are then washed over the side of the pan, leaving the gold behind. Once a placer deposit is located by gold panning, the miner usually shifts to equipment that can treat volumes of sand and gravel more quickly and efficiently.

Sluice box

The same principle may be employed on a larger scale by constructing a short sluice box, with barriers along the bottom called riffles to trap the heavier gold particles as water washes them and the other material along the box. This method better suits excavation with shovels or similar implements to feed ore into the device. Sluice boxes can be as short as a few feet, or more than ten feet (a common term for one that is over six feet +/- is a “Long Tom”).

Similar in principle to a sluice is a *rocker*, a cradle-like piece of equipment that could be rocked to sift sands through screens, which was introduced by Chinese miners in British Columbia and Australia, where the practice was referred to as “rocking the golden baby”. Another Chinese technique was the use of blankets to filter sand and gravels, catching fine gold in the fabric’s weave, then burning the blankets to smelt the gold. Chinese were noted for the thoroughness of their placer extraction techniques, which included hand-washing of individual rocks as well as the complete displacement of streambeds and advanced flume and ditching techniques which became copied by other miners.

Trommel

A trommel is composed of a slightly-inclined rotating metal tube (the ‘scrubber section’) with a screen at its discharge end. Lifter bars, sometimes in the form of bolted in angle iron, are attached to the interior of the scrubber section. The ore is fed into the elevated end of the trommel. Water, often under pressure, is provided to the scrubber and screen sections and the combination of water and mechanical action frees the valuable minerals from the ore. The mineral bearing ore that passes through the screen is then further concentrated in smaller devices such as sluices and jigs. The larger pieces of ore that do not pass through the screen can be carried to a waste stack by a conveyor.

Deposits

A well protected area from the flow of water is a great location to find gold (McCracken). Gold is very heavy and is often found in a stream bed (McCracken). Many different gold deposits are dealt with in different ways. Placer deposits attract many prospectors because their costs are very low (Lawler). There are many different places gold could be placed, such as a residual, alluvial, and a bench deposit.

Residual

Residual deposits are more common where there has been weathering on rocks and where there hasn't been water (Lawler). They are deposits which have not been washed away yet or been moved. The residual usually lies at the site of the lode (McCracken). This type of deposit undergoes rock weathering ("Types of Placers,").

Alluvial

Alluvial or eluvial deposits sometimes have the largest gold deposit and are very common (Lawler). This deposit is created when a force of nature moves or washes the gold away, but it doesn't go into a stream bed (McCracken). It contains pieces of ore that have been washed away from the lode. Alluvial deposits are the most common type of placer gold (McCracken). This type of deposit occurs mostly in valleys ("Types of Placers,").

Bench

Bench deposits are created when gold reaches a stream bed (McCracken). Gold accumulations in an old stream bed that are high are called bench deposits. They can be found on higher slopes that drain into valleys (Lawler). Dry stream beds (benches) can be situated far from other water sources and can sometimes be found on mountain tops. Today, many miners focus their activities on bench deposits.

Environmental Effects

Although not required, the process water may be continuously recycled and the ore from which the sought after minerals have been extracted ("the tailings") can be reclaimed. While these recycling and reclamation processes are more common in modern placer mining operations they are still not universally done.

In earlier times the process water was not generally recycled and the spent ore was not reclaimed. The remains of a Roman alluvial gold mine at Las Médulas are so spectacular as to justify the site being designated UNESCO World Heritage status.

The methods used by the Roman miners are fully described by Pliny the Elder in his work Naturalis Historia published in about 77 AD. The author was a Procurator in the region and so probably witnessed large-scale hydraulic mining of the placer deposits there. He also added that the local lake Curacado had been heavily silted by the mining methods.

Environmental activists describe the hydraulic mining form of placer mining as environmentally destructive because of the large amounts of silt that it adds to previously clear running streams (also known as the "Dahlonega Method"). Most placer mines today use settling ponds, if only to ensure that they have sufficient water to run their sluicing operations.

In California, from 1853 to 1884, "hydraulicking" of placers removed an enormous amount of material from the gold fields, material that was carried downstream and raised the level of the Central Valley by some seven feet in some areas and settled in long bars up to 20 feet thick in parts of San Francisco Bay.

The process raised an opposition calling themselves the "Anti-Debris Association". In January 1884, a United States District Court banned the flushing of debris into streams, and the hydraulic mining mania in California's gold country came to an end.

Drift Mining

Drift mining is either the mining of a placer deposit by underground methods, or the working of coal seams accessed by adits driven into the surface outcrop of the coal bed. Drift is a more general mining term, meaning a near-horizontal passageway in a mine, following the bed (of coal, for instance) or vein of ore. A drift may or may not intersect the ground surface. This kind of mining is done when the rock or mineral is on the side of a hill.

Nome, Alaska

Drift mining methods were used extensively to mine placer deposits during the early years of the Nome mining district. During summer, surface deposits could be worked, but some placer deposits were buried too deeply for surface placering. In addition, water to wash the gold from the placers was not available in the winter. Many miners tunnelled into deep placer deposits, bringing out the high-grade gravels to be washed at the spring thaw. Most of the ground in Nome is permafrost. By drift mining, miners were able to recover much of the gold buried under the permafrost.

Gold at Nome was concentrated in three ancient beach lines, now inshore, above sea level, and buried under roughly fifty feet of permafrost overlain by two feet of tundra. Gold was usually found on top of either bedrock or "false bedrock," a layer of clay that occurred at the base of the beach or stream deposit.

The miners initially sank shafts to prospect for the pay streaks. This was done by building a fire atop the permafrost, and, each day as it melted, shovelling away the mud. The process would be continued down to either a pay streak or bedrock. When the gold was found, the drift mining began.

Once they had found a pay streak, miners would tunnel horizontally from the bottom of their prospect shaft, and follow the gold along the surface of the bedrock.

The tunnels, because the ground was frozen, would not cave in. Miners discovered old underground beach and river gravels rich with gold. Around 1900 the population of Nome was more than twenty thousand, many of them drift miners. Nome's gold fields, appearing untouched from the surface, are honeycombed with tunnels left by the gold rush drift miners. Today's miners, prospecting with modern drilling equipment, sometimes hit old drifts.

Underground Mining (hard rock)

Underground hard rock mining refers to various underground mining techniques used to excavate *hard* minerals, mainly those containing metals such as ore containing gold, silver, iron, copper, zinc, nickel and lead, but also involves using the same techniques for excavating ores of gems such as diamonds. In contrast soft rock mining refers to excavation of softer minerals such as salt, coal, or tar sands.

Mine Access Underground Access

Accessing underground ore can be achieved via a decline (ramp), inclined vertical shaft or adit.

- Declines can be a spiral tunnel which circles either the flank of the deposit or circles around the deposit. The decline begins with a box cut, which is the portal to the surface. Depending on the amount of overburden and quality of bedrock, a galvanized steel culvert may be required for safety purposes. They may also be started into the wall of an open cut mine.
- Shafts are vertical excavations sunk adjacent to an ore body. Shafts are sunk for ore bodies where haulage to surface via truck is not economical. Shaft haulage is more economical than truck haulage at depth, and a mine may have both a decline and a ramp.
- Adits are horizontal excavations into the side of a hill or mountain. They are used for horizontal or near-horizontal ore bodies where there is no need for a ramp or shaft.

Declines are often started from the side of the high wall of an open cut mine when the ore body is of a payable grade sufficient to support an underground mining operation but the strip ratio has become too great to support open cast extraction methods. They are also often built and maintained as an emergency safety access from the underground workings and a means of moving large equipment to the workings.

Ore Access

Levels are excavated horizontally off the decline or shaft to access the ore body. Stopes are then excavated perpendicular (or near perpendicular) to the level into the ore.

Development Mining vs. Production Mining

There are two principal phases of underground mining: development mining and production mining.

Development mining is composed of excavation almost entirely in (non-valuable) waste rock in order to gain access to the orebody. There are six steps in development mining: remove previously blasted material (muck out round), Scaling (removing any unstable slabs of rock hanging from the roof and sidewalls to protect workers and equipment from damage), support excavation, drill rock face, load explosives, and blast explosives.

Production mining is further broken down into two methods, long hole and short hole. Short hole mining is similar to development mining, except that it occurs in ore. There are several different methods of long hole mining. Typically long hole mining requires two excavations within the ore at different elevations below surface, (15 m – 30 m apart). Holes are drilled between the two excavations and loaded with explosives. The holes are blasted and the ore is removed from the bottom excavation.

Ventilation

One of the most important aspects of underground hard rock mining is ventilation. Ventilation is required to clear toxic fumes from blasting and removing exhaust fumes from diesel equipment. In deep hot mines ventilation is also required for cooling the workplace for miners. Ventilation raises are excavated to provide ventilation for the workplaces, and can be modified for use as emergency escape routes.

The primary sources of heat in underground hard rock mines are virgin rock temperature, machinery, auto compression, and fissure water. Other small contributing factors are human body heat and blasting.

Ground Support

Some means of support is required in order to maintain the stability of the openings that are excavated. This support comes in two forms, local support and area support.

Area Ground Support

Area ground support is used to prevent major ground failure. Holes are drilled into the back (ceiling) and walls and a long steel rod (or rock bolt) is installed to hold the ground together. There are three categories of rock bolt, differentiated by how they engage the host rock. They are:

Mechanical Bolts

- Point anchor bolts (or expansion shell bolts) are a common style of area ground support. A point anchor bolt is a metal bar between 20 mm – 25 mm in diameter, and between 1 m – 4 m long (the size is determined by the mine's engineering department). There is an expansion shell at the end of the bolt which is inserted into the hole. As the bolt is tightened by the installation drill the expansion shell expands and the bolt tightens holding the rock together. Mechanical bolts are considered temporary support as their lifespan is reduced by corrosion as they are not grouted.

Grouted Bolts

- Resin grouted rebar is used in areas which require more support than a point anchor bolt can give. The rebar used is of similar size as a point anchor bolt but does not have an expansion shell. Once the hole for the rebar is drilled, cartridges of epoxy resin are installed in the hole. The rebar bolt is installed after the resin and spun by the installation drill. This opens the resin cartridge and mixes it. Once the resin hardens the drill spinning tightens the rebar bolt holding the rock together. Resin grouted rebar is considered a permanent ground support with a lifespan of 20–30 years.
- Cable bolts are used to bind large masses of rock in the hanging wall and around large excavations. Cable bolts are much larger than standard rock bolts and rebar, usually between 10–25 metres long. Cable bolts are grouted with a cement grout..

Friction Bolts

- Friction stabilizer (frequently called by the genericized trademark *Split Set*) are much easier to install than mechanical bolts or grouted bolts. The bolt is hammered into the drill hole, which has a smaller diameter than the bolt. Pressure from the bolt on the wall holds the rock together. Friction stabilizers are particularly susceptible to corrosion and rust from water unless they are grouted. Once grouted the friction increases by a factor of 3-4.
- Swellex is similar to Friction stabilizers, except the bolt diameter is smaller than the hole diameter. High pressure water is injected into the bolt to expand the bolt diameter to hold the rock together. Like the friction stabilizer, swellex is poorly protected from corrosion and rust.

Local Ground Support

Local ground support is used to prevent smaller rocks from falling from the backs and walls. Not all excavations require local ground support.

- Welded Wire Mesh is a metal screen with 10 cm x 10 cm (4 inch) openings. It is held to the backs using point anchor bolts or resin grouted rebar.
- Shotcrete is fibre reinforced spray on concrete which coats the

backs and walls preventing smaller rocks from falling. Shotcrete thickness can be between 50 mm – 100 mm.

- Latex Membranes can be sprayed on the backs and walls similar to shotcrete, but in smaller amounts.

Stope and Retreat vs. Stope and fill

Stope and Retreat

Using this method, mining is planned to extract rock from the stopes without filling the voids; this allows the wall rocks to cave in to the extracted stope after all the ore has been removed. The stope is then sealed to prevent access.

Stope and fill

Where large bulk ore bodies are to be mined at great depth, or where leaving pillars of ore is uneconomical, the open stope is filled with backfill, which can be a cement and rock mixture, a cement and sand mixture or a cement and tailings mixture. This method is popular as the refilled stopes provide support for the adjacent stopes, allowing total extraction of economic resources.

Mining Methods

Selective Mining Methods

- Cut and Fill mining is a method of short hole mining used in steeply dipping or irregular ore zones, in particular where the hanging wall limits the use of long hole methods. The ore is mined in horizontal or slightly inclined slices, and then filled with waste rock, sand or tailings. Either fill option may be consolidated with concrete, or left unconsolidated. Cut and fill mining is an expensive but selective method, with low ore loss and dilution.
- Drift and Fill is similar to cut and fill, except it is used in ore zones which are wider than the method of drifting will allow to be mined. In this case the first drift is developed in the ore, and is backfilled using consolidated fill. The second drift is driven adjacent to the first drift. This carries on until the ore zone is mined out to its full width, at which time the second cut is started atop of the first cut.
- Shrinkage Stoping is a short hole mining method which is suitable for steeply dipping orebodies. The method is similar

to cut and fill mining with the exception that after being blasted, broken ore is left in the stope where it is used to support the surrounding rock and as a platform from which to work. Only enough ore is removed from the stope to allow for drilling and blasting the next slice. The stope is emptied when all of the ore has been blasted. Although it is very selective and allows for low dilution, since the most of the ore stays in the stope until mining is completed there is a delayed return on capital investments.

- Room and Pillar mining : Room and pillar mining is commonly done in flat or gently dipping bedded ore bodies. Pillars are left in place in a regular pattern while the rooms are mined out. In many room and pillar mines, the pillars are taken out starting at the farthest point from the stope access, allowing the roof to collapse and fill in the stope. This allows for greater recovery as less ore is left behind in pillars.

Bulk Mining Methods

- Block Caving is used to mine massive steeply dipping orebodies (typically low grade) with high friability. An undercut with haulage access is driven under the orebody, with "drawbells" excavated between the top of the haulage level and the bottom of the undercut. The drawbells serve as a place for caving rock to fall into. The orebody is drilled and blasted above the undercut, and the ore is removed via the haulage access. Due to the friability of the orebody the ore above the first blast caves and falls into the drawbells. As ore is removed from the drawbells the orebody caves in providing a steady stream of ore. If caving stops and removal of ore from the drawbells continues, a large void may form, resulting in the potential for a sudden and massive collapse and potentially catastrophic windblast throughout the mine. Where caving does continue, the ground surface may collapse into a surface depression such as those at the Climax and Henderson molybdenum mines in Colorado. Such a configuration is one of several to which miners apply the term "glory hole".

Orebodies that do not cave readily are sometimes preconditioned by hydraulic fracturing, blasting, or by a combination of both. Hydraulic fracturing has been applied to preconditioning strong roof rock over coal longwall panels, and to inducing caving in both coal and hard rock mines.

Ore Removal

In mines which use rubber tired equipment for coarse ore removal, the ore (or "muck") is removed from the stope (referred to as "mucked out" or "bogged") using centre articulated vehicles (referred to as boggers or LHD [short for Load, Haul, Dump]). These pieces of equipment may operate using diesel or electric engines and resemble a low-profile front end loader.

The ore is then dumped into a truck to be hauled to the surface (in shallower mines). In deeper mines the ore is dumped down an ore pass (a vertical or near vertical excavation) where it falls to a collection level. On the collection level, it may receive primary crushing via jaw or cone crusher.

The ore is then moved by conveyor belts, trucks or occasionally trains to the shaft to be hoisted to the surface in buckets or skips and emptied into bins beneath the surface headframe for transport to the mill.

In some cases the underground primary crusher feeds an inclined conveyor belt which delivers ore via an incline shaft direct to the surface. The ore is fed down ore passes, with mining equipment accessing the ore body via a decline from surface.

Deepest Mines

- The deepest mines in the world are the TauTona (Western Deep Levels) and Savuka gold mines in the Witwatersrand region of South Africa, which are currently working at depths exceeding 3,900 m (12,800 ft). There are plans to extend Mponeng mine, a sister mine to TauTona, down to 4,500 m (14,800 ft) in the coming years.
- The deepest hard rock mine in North America is Agnico-Eagle's LaRonde mine, which mines gold, zinc, copper and silver ores roughly 45 km (28 mi) east of Rouyn-Noranda in Cadillac, Quebec. LaRonde's Penna shaft (#3 shaft) is believed to be the deepest single lift shaft in the Western Hemisphere. The new #4 shaft bottoms out at over 3,000 m (9,800 ft) down. Their LaRonde mine expansion sees open stopes down to a depth of over 3,000 m (9,800 ft), the deepest longhole open stopes in the world.

- The deepest hard rock mines in Australia are the copper and zinc lead mines in Mount Isa, Queensland at 1,800 m (5,900 ft).
- The deepest platinum-palladium mines in the world are on the Merensky Reef, in South Africa, with a resource of 203 million Troy ounces, currently worked to approximately 2,200 m (7,200 ft) depth.
- The harshest conditions for hard rock mining are in the Witwatersrand area of South Africa, where workers toil in temperatures of up to 45 °C (113 °F). However, massive refrigeration plants are used to bring the air temperature down to around 28 °C (82 °F).

Underground Mining (soft rock)

Underground mining (soft rock) refers to a group of underground mining techniques used to extract coal, oil shale and other minerals or geological materials from sedimentary ("soft") rocks. Because deposits in sedimentary rocks are commonly layered and relatively less hard, the mining methods used differ from those used to mine deposits in igneous or metamorphic rocks. Underground mining techniques also differ greatly from those of surface mining.

Methods

- Longwall mining-A set of longwall mining equipment consists of a coal shearer mounted on conveyor operating underneath a series of self-advancing hydraulic roof supports. Almost the entire process can be automated. Longwall mining machines are typically 150-250 metres in width and 1.5 to 3 metres high. Longwall miners extract "panels"-rectangular blocks of coal as wide as the face the equipment is installed in, and as long as several kilometres. Powerful mechanical coal cutters (shearers) cut coal from the face, which falls onto an armoured face conveyor for removal. Longwalls can advance into an area of coal, or more commonly, retreat back between development tunnels (called "gateroads") As a longwall miner retreats back along a panel, the roof behind the supports is allowed to collapse in a planned and controlled manner.
- Room-and-pillar mining or continuous mining-Room and pillar mining is commonly done in flat or gently dipping bedded ores. Pillars are left in place in a regular pattern while the rooms are mined out. In many room and pillar mines, the pillars are

taken out, starting at the farthest point from the mine haulage exit, retreating, and letting the roof come down upon the floor. Room and pillar methods are well adapted to mechanization, and are used in deposits such as coal, potash, phosphate, salt, oil shale, and bedded uranium ores.

- Blast mining – An older practice of coal mining that uses explosives such as dynamite to break up the coal seam, after which the coal is gathered and loaded onto shuttle cars or conveyors for removal to a central loading area. This process consists of a series of operations that begins with "cutting" the coalbed so it will break easily when blasted with explosives. This type of mining accounts for less than 5% of total underground production in the U.S. today.
- Shortwall mining– A coal mining method that accounts for less than 1% of deep coal production, shortwall involves the use of a continuous mining machine with moveable roof supports, similar to longwall. The continuous miner shears coal panels 150–200 feet wide and more than a half-mile long, depending on other things like the strata of the Earth and the transverse waves.
- Coal Skimming While no longer in general use, because of the massive amount of water needed and environmental damage there of, in the late 1930s DuPont developed a method that was much faster and less labour intensive than previous methods to separate the lighter coal from the mining refuse (e.g. slate) called "coal skimming" or the "sink and float method".

Mine Shorthand

The number sign, or hash sign (#) is often used as shorthand to denote shaft or seam, as in 4# (4 shaft or seam depending on context).

Borehole Mining

Borehole Mining (BHM) (also referred to as jet-boring) is a remote operated method of extracting (mining) mineral resources through boreholes by means of high pressure water jets. This process can be carried-out from land surface, open pit floor, underground mine or floating platform or vessel through pre-drilled boreholes.

The Process

1. A borehole is drilled from land surface to a desired depth, where then actual borehole mining will take place.

2. A casing column is lowered down the hole. Since BHM takes place in an "open hole", the casing shoe is located just above the upper border of the production interval (ore body) leaving the rest open.
3. The BHM tool is lowered into the hole.

Description of a BHM tool

The tool consists of at least two concentric pipes which are forming two hydraulic channels-one for pumping down a high-pressure working agent (water) and second for delivering pregnant slurry back to the surface. A BHM tool usually has (down-up): an eductor (waterjet pump) section, a hydromonitor section, an extension section and a hub, connecting it all to a drill pipe string. This string extends the tool up to the surface. Above the surface, the tool has a swivel allowing its suspension and rotation in a hole, and also connections to the working agent supply (pump station) and a slurry collector. A drill rig is normally required to operate a BHM tool.

The tool is lowered into a well until the hydromonitor reaches the required depth where the actual borehole mining is started. Then the high-pressure water is pumped down and receives back productive slurry. In the collecting pond or tank, slurry is separated and clarified water is pumped down for re-circulating.

While extracting of material, different shape underground caverns could be created. Their shapes depend on the BHM tool manipulation while mining, which obviously consist of the tool rotation, sliding it up and down and combination of these two movements. Borehole mining is applied from vertical, horizontal and deviated wells.

Advantages of BHM

The main advantages of BHM include its low capital cost, mobility, selectivity, ability to work in hazardous and dangerous conditions and low environmental impact. The method has been used in mining of such natural resources and industrial materials as: uranium, iron ore, quartz sand, gravel, coal, poly-metallic ores, phosphate, gold, diamonds, rare earths, amber and several more. Borehole mining is also used in exploration, oil, gas and water stimulation, underground storage construction and drainage.

Longwall Mining

Longwall mining is a form of underground coal mining where a long wall of coal is mined in a single slice (typically 1–2 m thick). The longwall *panel* (the block of coal that is being mined) is typically 3–4 km long and 250–400 m wide.

History

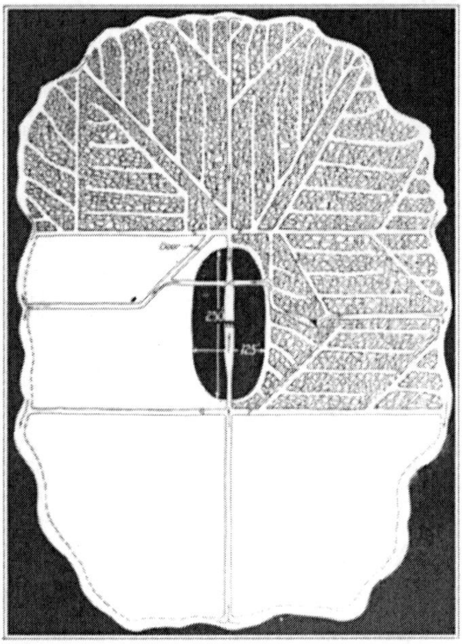

Figure 1 : Plan of Longwall Mine before Conveyors, Hoist is at the Centre of the Central Pillar.

The basic idea of longwall mining was developed in England in the late 17th century. Miners would undercut the coal along the width of the coal face, removing coal as it fell, and using wooden props to control the fall of the roof behind the face. this was known as the *Shropshire method* of mining.

While the technology has changed considerably, the basic idea remains the same, to remove essentially all of the coal from a broad coal face and allow the roof and overlying rock to collapse into the void behind, while maintaining a safe working space along the face for the miners.

Starting around 1900, mechanization was applied to this method. By 1940, some referred to longwall mining as "the conveyor method" of mining, after the most prominent piece of machinery involved. Unlike earlier longwall mining, the use of a conveyor belt parallel to the coal face forced the face to be developed along a straight line. The only other machinery used was an electric cutter to undercut the coal face and electric drills for blasting to drop the face. Once dropped, manual labour was used to load coal onto the conveyor parallel to the face and to place wooden roof props to control the fall of the roof.

Such low-technology longwall mines continued in operation into the 1970s. The best known example was the New Gladstone Mine near Centerville, Iowa "one of the last advancing longwall mines in the United States". This longwall mine did not even use a conveyor belt, but relied on ponies to haul coal tubs from the face to the slope where a hoist hauled the tubs to the surface.

Longwall mining has been extensively used as the final stage in mining old room and pillar mines. In this context, longwall mining can be classified as a form of retreat mining.

Layout

Gate roads are driven to the back of each panel before longwall mining begins. The gate road along one side of the block is called the *maingate* or *headgate*; the road on the other side is called the *tailgate*. Where the thickness of the coal allows, these gate roads have been previously developed by continuous miner units, as the longwall itself is not capable of the initial development. In thinner seams the advancing longwall mining method may be used. In this system the gate roads are formed as the coal face advances.

Only the maingate road is formed in advance of the face. The tailgate road is formed behind the coal face by removing the stone above coal height to form a roadway that is high enough to travel in. The end of the block that includes the longwall equipment is called the face. The other end of the block is usually one of the main travel roads of the mine. The cavity behind the longwall is called the *goaf*, *goff* or *gob*.

Ventilation

Fresh air travels up the main gate, across the face, and then down the tail gate. Once past the face the air is no longer fresh air, but return air carrying away coal dust and mine gases such as methane, carbon dioxide, depending on the geology of the coal. Return air is extracted by ventilation fans mounted on the surface. A series of seals are erected as mining progresses to maintain goaf gas levels.

Typically to avoid coal in the goaf spontaneously combusting, goaf gases are allowed to build up so as to exclude oxygen from the goafed area. This means that there is an explosive goaf fringe between the face and the goaf at all times requiring constant monitoring.

Equipment

A number of hydraulic jacks, called *powered roof supports*, *chocks* or *shields*, which are typically 1.75 m wide and placed in a long line, side by side for up to 400 m in length in order to support the roof of the coalface. An individual chock can weigh 30–40 tonnes, extend to a maximum cutting height of up to 6 m and have yield rating of 1000–1250 tonnes each, and hydraulically advance itself 1 m at a time.

The coal is cut from the coalface by a machine called the *shearer* (*power loader*). This machine can weigh 75–120 tonnes typically and comprises a main body, housing the electrical functions, the tractive motive units to move the shearer along the coalface and pumping units (to power both hydraulic and water functions). At either end of the main body are fitted the ranging arms which can be ranged vertically up down by means of hydraulic rams, and onto which are mounted the shearer cutting drums which are fitted 40–60 cutting picks. Within the ranging arms are housed very powerful electric motors (typically up to 850 kW) which transfer their power through a series of lay gears within the body the arms to the drum mounting locations at the extreme ends of the ranging arms where the cutting drums are. The cutting drums are rotated at a speed of 20–50 revs/min to cut the mineral from coal seam.

The shearer is carried along the length of the face on the *armoured face conveyor* (AFC); using a chain-less haulage system, which resembles a ruggedised rack and pinion system especially developed for mining. Before chainless haulage systems, a heavy duty chain was run the length of the coal face for the shearer to pull itself along. The shearer moves at a speed of 10–30 m/min depending on cutting conditions.

The AFC is placed in front of the powered roof supports, and the shearing action of the rotating drums cutting into the coal seam disintegrates the coal, this being loaded onto the AFC. The coal is removed from the coal face by a scraper chain conveyor to the main gate. Here it is loaded onto a network of conveyor belts for transport to the surface. At the main gate the coal is usually reduced in size in a crusher, and loaded onto the first conveyor belt by the *beam stage loader* (BSL).

As the shearer removes the coal, the AFC is snaked over behind the shearer and the powered roof supports move forward into the newly created cavity. As mining progresses and the entire longwall progresses through the seam, the goaf increases. This goaf collapses under the weight of the overlying strata. The strata approximately 2.5 times the thickness of the coal seam removed collapses and the beds above settle onto the collapsed goaf. This collapsing can lower surface height, causing problems like changing the course of rivers and severely damage building foundations.

Comparison with Room and Pillar Method

Longwall and room and pillar methods of mining can both be used for mining suitable underground coal seams. Longwall has better resource recovery (about 80% compared with about 60% for room and pillar method, fewer roof support consumables are needed, higher volume coal clearance systems, minimal manual handling and safety of the miners is enhanced by the fact that they are always under the hydraulic roof supports when they are extracting coal.

Subsidence

Subsidence is largely immediate, allowing for better planning and more accountability by the mining company. There have been cases of surface subsidence altering the landscape above the mines. At Newstan Colliery in New South Wales, Australia "the surface has dropped by as much as five metres in places" above a multi level mine. In some cases the subsidence causes damage to natural features such as drainage to water courses or man-made structures such as roads and buildings. "Douglas Park Drive was closed for four weeks because longwall panels ... destabilised the road. In 2000, the State Government stopped mining when it came within 600 metres from the twin bridges. A year later there were reports of 40-centimetre gaps appearing in the road, and the bridge had to be jacked sideways to realign it."

A 2005 geotechnical report commissioned by the NSW RTA warns that "subsidence could happen suddenly and occur over many years".

Retreat Mining

Retreat mining is a term used to reference the final phase of an underground mining technique known as room and pillar mining. This involves excavating a room or chamber while leaving behind pillars of material for support. This excavation is carried out in a pattern advancing away from the entrance of a mine. Once a deposit has been exhausted using this method, the pillars that were left behind initially are removed, or 'pulled', retreating back towards the mine's entrance. After the pillars are removed, the roof (or back) is allowed to collapse behind the mining area. Pillar removal must occur in a very precise order in order to reduce the risks to workers, due to the high stresses placed on the remaining pillars by the abutment stresses of the caving ground.

Retreat mining is the term commonly referred to as the mining method employed by the Crandall Canyon Mine and is believed, by some, to be a factor in the August 2007 collapse which trapped six miners. On November 21, 2007, the mine was sealed, entombing the deceased miners. The bureau of land management says retreat mining was being used. Robert E. Murray, CEO of Murray Energy (owner of the mine) says the technique was not being used at Crandall Canyon.

Chapter 2

Mineralogy

Mineralogy is the study of chemistry, crystal structure, and physical (including optical) properties of minerals. Specific studies within mineralogy include the processes of mineral origin and formation, classification of minerals, their geographical distribution, as well as their utilization.

History of Mineralogy

Early writing on mineralogy, especially on gemstones, comes from ancient Babylonia, the ancient Greco-Roman world, ancient and medieval China, and Sanskrit texts from ancient India. Books on the subject included the Naturalis Historia of Pliny the Elder which not only described many different minerals but also explained many of their properties. The German Renaissance specialist Georgius Agricola wrote works such as *De re metallica* (*On Metals*, 1556) and *De Natura Fossilium* (*On the Nature of Rocks*, 1546) which begin the scientific approach to the subject. Systematic scientific studies of minerals and rocks developed in post-Renaissance Europe. The modern study of mineralogy was founded on the principles of crystallography and microscopic study of rock sections with the invention of the microscope in the 17th century.

Europe and the Middle East

The ancient Greek writers Aristotle (384–322 BC) and Theophrastus (370-285 BC) were the first in the Western tradition to write of minerals and their properties, as well as metaphysical explanations for them. The Greek philosopher Aristotle wrote his *Meteorologica*, and in it theorized that all the known substances were composed of water, air, earth, and fire, with the properties of dryness, dampness, heat, and cold. The Greek philosopher and botanist

Theophrastus wrote his *De Mineralibus*, which accepted Aristotle's view, and divided minerals into two categories: those affected by heat and those affected by dampness.

The metaphysical emanation and exhalation (*anathumiaseis*) theory of the Greek philosopher Aristotle included early speculation on earth sciences including mineralogy. According to his theory, while metals were supposed to be congealed by means of moist exhalation, dry gaseous exhalation (*pneumatodestera*) was the efficient material cause of minerals found in the Earth's soil.

He postulated these ideas by using the examples of moisture on the surface of the earth (a moist vapour 'potentially like water'), while the other was from the earth itself, pertaining to the attributes of hot, dry, smoky, and highly combustible ('potentially like fire'). Aristotle's metaphysical theory from times of antiquity had wide-ranging influence on similar theory found in later medieval Europe, as the historian Berthelot notes:

The theory of exhalations was the point of departure for later ideas on the generation of metals in the earth, which we meet with Proclus, and which reigned throughout the middle ages.

Ancient Greek terminology of minerals has also stuck through the ages with widespread usage in modern times.

For example, the Greek word asbestos (meaning 'inextinguishable', or 'unquenchable'), for the unusual mineral known today containing fibrous structure. The ancient historians Strabo (63 BC-19 AD) and Pliny the Elder (23-79 AD) both wrote of asbestos, its qualities, and its origins, with the Hellenistic belief that it was of a type of vegetable. Pliny the Elder listed it as a mineral common in India, while the historian Yu Huan (239-265 AD) of China listed this 'fireproof cloth' as a product of ancient Rome or Arabia (Chinese: Daqin). Although documentation of these minerals in ancient times does not fit the manner of modern scientific classification, there was nonetheless extensive written work on early mineralogy.

Pliny the Elder

For example, Pliny devoted 5 entire volumes of his work Naturalis Historia (77 AD) to the classification of "earths, metals, stones, and gems". He not only describes many minerals not known to Theophrastus, but discusses their applications and properties. He is the first to correctly recognise the origin of amber for example, as the fossilized remnant of tree resin from the observation of insects trapped in some samples. He laid the basis of crystallography by discussing crystal habit, especially the octahedral shape of diamond. His discussion of mining methods is unrivalled in the ancient world, and includes, for example, an eye-witness account of gold mining in northern Spain, an account which is fully confirmed by modern research.

However, before the more definitive foundational works on mineralogy in the 16th century, the ancients recognized no more than roughly 350 minerals to list and describe.

Jabir and Avicenna

With philosophers such as Proclus, the theory of Neoplatonism also spread to the Islamic world during the Middle Ages, providing a basis for metaphyiscal ideas on mineralogy in the medieval Middle East as well. The medieval Islamic scientists expanded upon this as well, including the Persian scientist Ibn Sina (980-1037 AD), also known as *Avicenna*, who rejected alchemy and the earlier notion of Greek metaphysics that metallic and other elements could be transformed into one another.

However, what was largely accurate of the ancient Greek and medieval metaphysical ideas on mineralogy was the slow chemical change in composition of the Earth's crust. There was also the Islamic alchemist and scientist Jabir ibn Hayyan (721-815 AD), who was the first to bring the experimental method into alchemy. Aided by Greek mathematics and Islamic mathematics, he discovered the syntheses for hydrochloric acid, nitric acid, distillation and crystallization (the latter two being essential for the understanding of modern mineralogy).

Georgius Agricola, Father of Mineralogy

In the early 16th century AD, the writings of the German scientist Georg Bauer, pen-name Georgius Agricola (1494-1555 AD), in his *Bermannus, sive de re metallica dialogus* (1530) is considered to be the official establishment of mineralogy in the modern sense of its study. He wrote the treatise while working as a town physician and making observations in Joachimsthal, which was then a centre for mining and metallurgic smelting industries. In 1544, he published his written work *De ortu et causis subterraneorum*, which is considered to be the foundational work of modern physical geology. In it (much like Ibn Sina) he heavily criticized the theories laid out by the ancient Greeks such as Aristotle. His work on mineralogy and metallurgy continued with the publication of *De veteribus et novis metallis* in 1546, and culminated in his best known works, the *De re metallica* of 1556. It was an impressive work outlining applications of mining, refining, and smelting metals, alongside discussions on geology of ore bodies, surveying, mine construction, and ventilation. He praises Pliny the Elder for his pioneering work Naturalis Historia and makes extensive references to his discussion of minerals and mining methods. For the next two centuries this written work remained the authoritative text on mining in Europe.

Agricola had many various theories on mineralogy based on empirical observation, including understanding of the concept of ore channels that were formed by the circulation of ground waters ('succi') in fissures subsequent to the deposition of the surrounding rocks. The medieval Chinese previously had conceptions of this as well.

For his works, Agricola is posthumously known as the "Father of Mineralogy".

After the foundational work written by Agricola, it is widely agreed by the scientific community that the *Gemmarum et Lapidum Historia* of Anselmus de Boodt (1550–1632) of Bruges is the first definitive work of modern mineralogy. The German mining chemist J.F. Henckel wrote his *Flora Saturnisans* of 1760, which was the first treatise in Europe to deal with geobotanical minerals, although the Chinese had mentioned this in earlier treatises of 1421 and 1664. In addition, the Chinese writer Du Wan made clear references to weathering and erosion processes in his *Yun Lin Shi Pu* of 1133, long before Agricola's work of 1546.

China and the Far East

In ancient China, the oldest literary listing of minerals dates back to at least the 4th century BC, with the *Ji Ni Zi* book listing twenty four of them. Chinese ideas of metaphysical mineralogy span back to at least the ancient Han Dynasty (202 BC-220 AD). From the 2nd century BC text of the *Huai Nan Zi*, the Chinese used ideological Taoist terms to describe meteorology, precipitation, different types of minerals, metallurgy, and alchemy.

Although the understanding of these concepts in Han times was Taoist in nature, the theories proposed were similar to the Aristotelian theory of mineralogical exhalations. By 122 BC, the Chinese had thus formulated the theory for metamorphosis of minerals, although it is noted by historians such as Dubs that the tradition of alchemical-mineralogical Chinese doctrine stems back to the School of Naturalists headed by the philosopher Zou Yan (305 BC-240 BC). Within the broad categories of rocks and stones (shi) and metals and alloys (jin), by Han times the Chinese had hundreds (if not thousands) of listed types of stones and minerals, along with theories for how they were formed.

In the 5th century AD, Prince Qian Ping Wang of the Liu Song Dynasty wrote in the encyclopedia *Tai-ping Yu Lan* (circa 444 AD, from the lost book *Dian Shu*, or *Management of all Techniques*):

The most precious things in the world are stored in the innermost regions of all. For example, there is orpiment. After a thousand years it changes into realgar. After another thousand years the realgar becomes transformed into yellow gold.

In ancient and medieval China, mineralogy became firmly tied to empirical observations in pharmaceutics and medicine. For example, the famous horologist and mechanical engineer Su Song (1020-1101 AD) of the Song Dynasty (960-1279 AD) wrote of mineralogy and pharmacology in his *Ben Cao Tu Jing* of 1070. In it he created a systematic approach to listing various different minerals and their use in medicinal concoctions, such as all the variously known forms of mica that could be used to cure various ills through digestion. Su Song also wrote of the subconchoidal fracture of native cinnabar, signs of ore beds, and provided description on crystal form. Similar to the ore channels formed by circulation of ground water with the German scientist Agricola, Su Song made similar statements concerning copper carbonate, as did the earlier *Ri Hua Ben Cao* of 970 AD with copper sulfate.

The Yuan Dynasty scientist Zhang Si-xiao (died 1332 AD) provided a groundbreaking treatise on the conception of ore beds from the circulation of ground waters and rock fissures, two centuries before Georgius Agricola would come to similar conclusions. In his *Suo-Nan Wen Ji*, he applies this theory in describing the deposition of minerals by evaporation of (or precipitation from) ground waters in ore channels.

In addition to alchemical theory posed above, later Chinese writers such as the Ming Dynasty physician Li Shizhen (1518-1593 AD) wrote of mineralogy in similar terms of Aristotle's metaphysical theory, as the latter wrote in his pharmaceutical treatise *Bìnclo Gângmù*. Another figure from the Ming era, the famous geographer Xu Xiake (1587–1641) wrote of mineral beds and mica schists in his treatise. However, while European literature on mineralogy became wide and varied, the writers of the Ming and Qing dynasties wrote little of the subject (even compared to Chinese of the earlier Song era). The only other works from these two eras worth mentioning were the *Shi Pin* (Hierarchy of Stones) of Yu Jun in 1617, the *Guai Shi Lu* (Strange Rocks) of Song Luo in 1665, and the *Guan Shi Lu* (On Looking at Stones) in 1668.

Theories of Shen Kuo

The medieval Chinese Song Dynasty statesman and scientist Shen Kuo (1031-1095 AD) wrote of his land formation theory involving concepts of mineralogy.

In his *Meng Xi Bi Tan*, Shen formulated a hypothesis for the process of land formation (geomorphology); based on his observation of marine fossil shells in a geological stratum in the Taihang Mountains hundreds of miles from the Pacific Ocean. He inferred that the land was formed by erosion of the mountains and by deposition of silt, and described soil erosion, sedimentation and uplift.

In an earlier work of his (circa 1080), he wrote of a curious fossil of a sea-orientated creature found far inland. It is also of interest to note that the contemporary author of the *Xi Chi Cong Yu* attributed the idea of particular places under the sea where serpents and crabs were petrified to one Wang Jinchen. With Shen Kuo's writing of the discovery of fossils, he formulated a hypothesis for the shifting of geographical climates throughout time.

This was due to hundreds of petrified bamboos found underground in the dry climate of northern China, once an enormous landslide upon the bank of a river revealed them. Shen theorized that in pre-historic times, the climate of Yanzhou must have been very rainy and humid like southern China, where bamboos are suitable to grow.

In a similar way, the historian Joseph Needham likened Shen's account with the Scottish scientist Roderick Murchison (1792–1871), who was inspired to become a geologist after observing a providential landslide.

In addition, Shen's description of sedimentary deposition predated that of James Hutton, who wrote his groundbreaking work in 1802 (considered the foundation of modern geology). The influential philosopher Zhu Xi (1130–1200) wrote of this curious natural phenomena of fossils as well, and was known to have read the works of Shen Kuo. In comparison, the first mentioning of fossils found in the West was made nearly two centuries later with Louis IX of France in 1253 AD, who discovered fossils of marine animals (as recorded in Joinville's records of 1309 AD).

Modern Mineralogy

Historically, mineralogy was heavily concerned with taxonomy of the rock-forming minerals; to this end, the International Mineralogical Association is an organization whose members represent mineralogists in individual countries.

Its activities include managing the naming of minerals (via the Commission of New Minerals and Mineral Names), location of known minerals, etc. As of 2004 there are over 4,000 species of mineral recognized by the IMA. Of these, perhaps 150 can be called "common," another 50 are "occasional," and the rest are "rare" to "extremely rare."

More recently, driven by advances in experimental technique (such as neutron diffraction) and available computational power, the latter of which has enabled extremely accurate atomic-scale simulations of the behaviour of crystals, the science has branched out to consider more general problems in the fields of inorganic chemistry and solid-state physics.

It, however, retains a focus on the crystal structures commonly encountered in rock-forming minerals (such as the perovskites, clay minerals and framework silicates).

In particular, the field has made great advances in the understanding of the relationship between the atomic-scale structure of minerals and their function; in nature, prominent examples would be accurate measurement and prediction of the elastic properties of minerals, which has led to new insight into seismological behaviour of rocks and depth-related discontinuities in seismograms of the Earth's mantle.

To this end, in their focus on the connection between atomic-scale phenomena and macroscopic properties, the mineral sciences (as they are now commonly known) display perhaps more of an overlap with materials science than any other discipline.

Physical Mineralogy

Physical mineralogy is the specific focus on physical attributes of minerals. Description of physical attributes is the simplest way to identify, classify, and categorize minerals, and they include:

Crystal Structure

In mineralogy and crystallography, crystal structure is a unique arrangement of atoms or molecules in a crystalline liquid or solid. A crystal structure is composed of a pattern, a set of atoms arranged in a particular way, and a lattice exhibiting long-range order and symmetry. Patterns are located upon the points of a lattice, which is an array of points repeating periodically in three dimensions. The points can be thought of as forming identical tiny boxes, called unit cells, that fill the space of the lattice. The lengths of the edges of a unit cell and the angles between them are called the *lattice parameters.* The symmetry properties of the crystal are embodied in its space group.

A crystal's structure and symmetry play a role in determining many of its physical properties, such as cleavage, electronic band structure, and optical transparency.

Unit Cell

The crystal structure of a material or the arrangement of atoms within a given type of crystal structure can be described in terms of its unit cell. The unit cell is a small box containing one or more atoms, a spatial arrangement of atoms. The unit cells stacked in three-dimensional space describe the bulk arrangement of atoms of the crystal. The crystal structure has a three-dimensional shape. The unit cell is given by its lattice parameters, which are the length of the cell edges and the angles between them, while the positions of the atoms inside the unit cell are described by the set of atomic positions (x_i, y_i, z_i) measured from a lattice point.

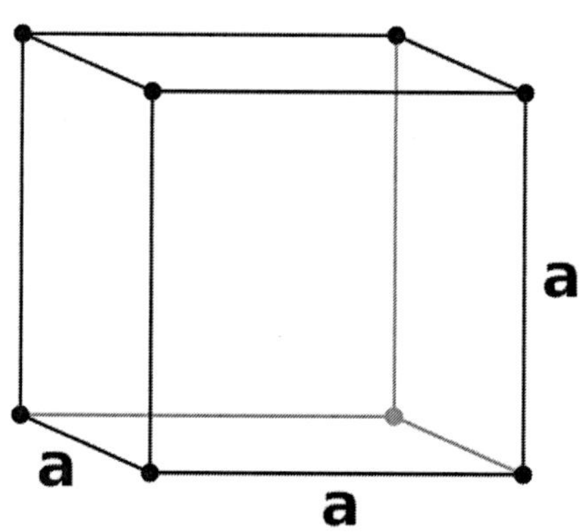

Figure 1 : Simple Cubic (P)

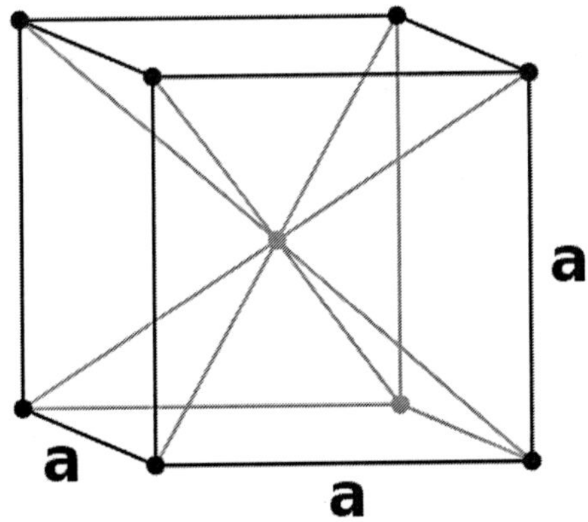

Figure 2 : Body-centred Cubic (I)

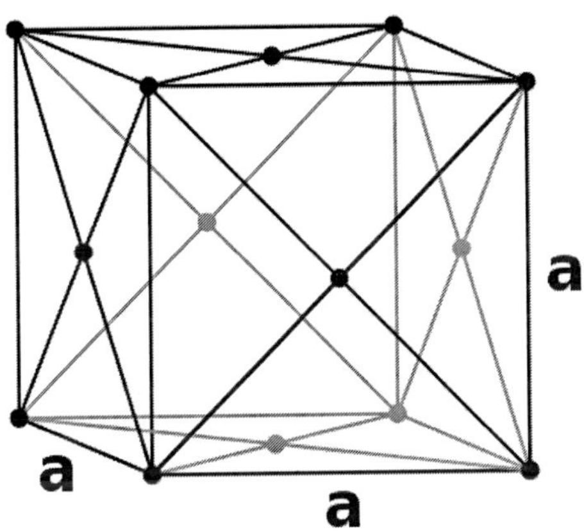

Figure 3 : Face-centred Cubic (F)

Miller Indices

Vectors and atomic planes in a crystal lattice can be described by a three-value Miller index notation (˘*!mn*). The ˘*!*, *m*, and *n* directional indices are separated by 90°, and are thus orthogonal. In fact, the ˘*!* component is mutually perpendicular to the *m* and *n* indices.

By definition, (˘*!mn*) denotes a plane that intercepts the three points a_1/–!, a_2/*m*, and a_3/*n*, or some multiple thereof.

That is, the Miller indices are proportional to the *inverses* of the intercepts of the plane with the unit cell (in the basis of the lattice vectors). If one or more of the indices is zero, it simply means that the planes do not intersect that axis (i.e., the intercept is "at infinity").

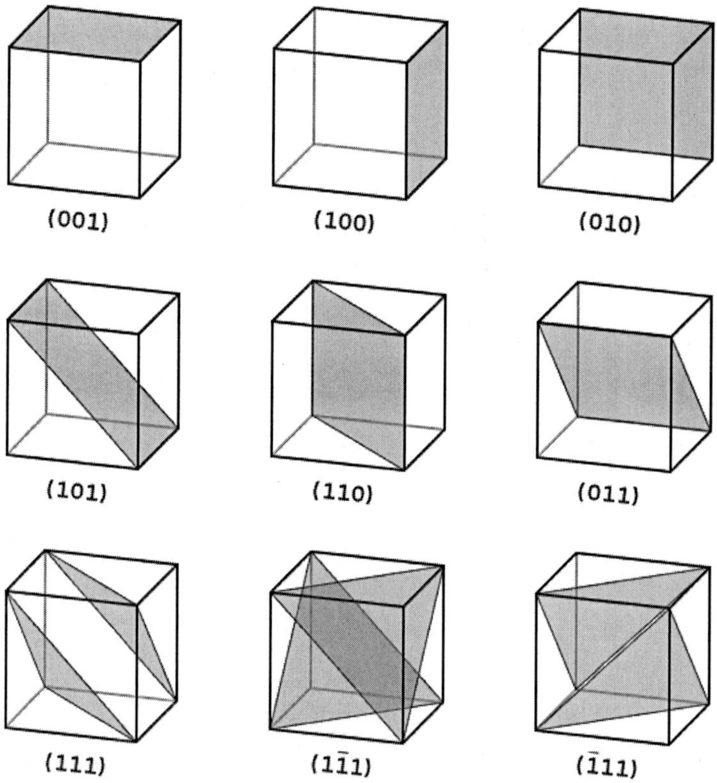

Figure 4: Planes with different Miller indices in cubic crystals

Considering only (˘!*mn*) planes intersecting one or more lattice points (the *lattice planes*), the perpendicular distance *d* between adjacent lattice planes is related to the (shortest) reciprocal lattice vector orthogonal to the planes by the formula:

$$d = 2\pi / |\mathbf{g}_{\ell mn}|$$

Planes and Directions

The crystallographic directions are fictitious lines linking nodes (atoms, ions or molecules) of a crystal. Likewise, the crystallographic planes are fictitious *planes* linking nodes. Some directions and planes have a higher density of nodes. These high density planes have an influence on the behaviour of the crystal as follows:

- Optical properties: Refractive index is directly related to density (or periodic density fluctuations).
- Adsorption and reactivity: Physical adsorption and chemical

reactions occur at or near surface atoms or molecules. These phenomena are thus sensitive to the density of nodes.

- Surface tension: The condensation of a material means that the atoms, ions or molecules are more stable if they are surrounded by other similar species. The surface tension of an interface thus varies according to the density on the surface.

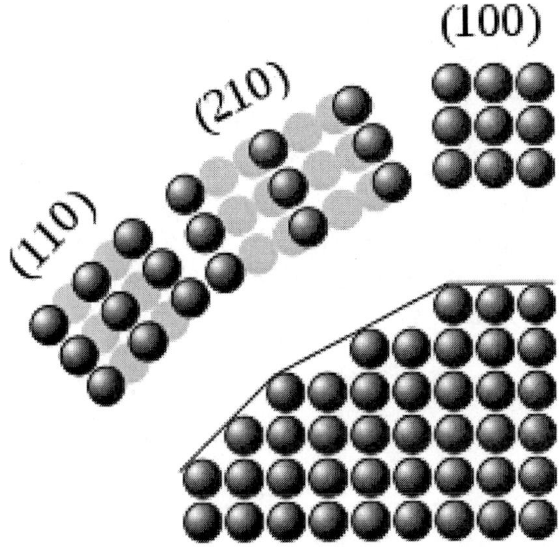

Figure 5: Dense Crystallographic Planes

- Microstructural defects: Pores and crystallites tend to have straight grain boundaries following higher density planes.
- Cleavage: This typically occurs preferentially parallel to higher density planes.
- Plastic deformation: Dislocation glide occurs preferentially parallel to higher density planes. The perturbation carried by the dislocation (Burgers vector) is along a dense direction. The shift of one node in a more dense direction requires a lesser distortion of the crystal lattice.

In the rhombohedral, hexagonal, and tetragonal systems, the basal plane is the plane perpendicular to the principal axis.

Cubic Structures

For the special case of simple cubic crystals, the lattice vectors are orthogonal and of equal length (usually denoted a); similarly for the reciprocal lattice. So, in this common case, the Miller indices (–!mn) and [–!mn] both simply denote normals/directions in Cartesian coordinates. For cubic crystals with lattice constant *a*, the spacing *d* between adjacent (–!mn) lattice planes is (from above):

$$d_{\ell mn} = \frac{a}{\sqrt{\ell^2 + m^2 + n^2}}$$

Because of the symmetry of cubic crystals, it is possible to change the place and sign of the integers and have equivalent directions and planes:

- Coordinates in *angle brackets* such as <100> denote a *family* of directions that are equivalent due to symmetry operations, such as [100], [010], [001] or the negative of any of those directions.
- Coordinates in *curly brackets* or *braces* such as {100} denote a family of plane normals that are equivalent due to symmetry operations, much the way angle brackets denote a family of directions.

For face-centred cubic (fcc) and body-centred cubic (bcc) lattices, the primitive lattice vectors are not orthogonal. However, in these cases the Miller indices are conventionally defined relative to the lattice vectors of the cubic supercell and hence are again simply the Cartesian directions.

Classification

The defining property of a crystal is its inherent symmetry, by which we mean that under certain 'operations' the crystal remains unchanged. For example, rotating the crystal 180° about a certain axis may result in an atomic configuration that is identical to the original configuration. The crystal is then said to have a twofold rotational symmetry about this axis. In addition to rotational symmetries like this, a crystal may have symmetries in the form of mirror planes and translational symmetries, and also the so-called "compound symmetries," which are a combination of translation and rotation/ mirror symmetries. A full classification of a crystal is achieved when all of these inherent symmetries of the crystal are identified.

Lattice Systems

These lattice systems are a grouping of crystal structures according to the axial system used to describe their lattice. Each lattice system consists of a set of three axes in a particular geometrical arrangement. There are seven lattice systems. They are similar to but not quite the same as the seven crystal systems and the six crystal families. The simplest and most symmetric, the cubic (or isometric) system, has the symmetry of a cube, that is, it exhibits four threefold rotational axes oriented at 109.5° (the tetrahedral angle) with respect to each other. These threefold axes lie along the body diagonals of the cube. The other six lattice systems, are hexagonal, tetragonal, rhombohedral (often confused with the trigonal crystal system), orthorhombic, monoclinic and triclinic.

Atomic Coordination

By considering the arrangement of atoms relative to each other, their coordination numbers (or number of nearest neighbours), interatomic distances, types of bonding, etc., it is possible to form a general view of the structures and alternative ways of visualizing then.

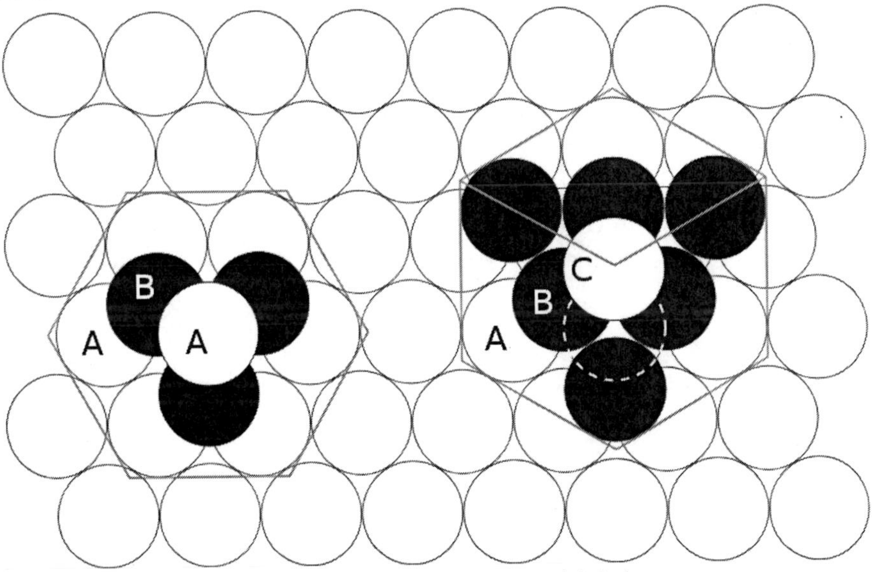

Figure 6: HCP lattice (left) and the fcc lattice (right).

Close Packing

The principles involved can be understood by considering the most efficient way of packing together equal-sized spheres and stacking close-packed atomic planes in three dimensions. For example, if plane A lies beneath plane B, there are two possible ways of placing an additional atom on top of layer B. If an additional layer was placed directly over plane A, this would give rise to the following series :

...ABABABAB....

This type of crystal structure is known as hexagonal close packing (hcp).

If however, all three planes are staggered relative to each other and it is not until the fourth layer is positioned directly over plane A that the sequence is repeated, then the following sequence arises:

...ABCABCABC...

This type of crystal structure is known as cubic close packing (ccp)

The unit cell of the ccp arrangement is the face-centred cubic (fcc) unit cell. This is not immediately obvious as the closely packed layers are parallel to the {111} planes of the fcc unit cell. There are four different orientations of the close-packed layers.

The packing efficiency could be worked out by calculating the total volume of the spheres and dividing that by the volume of the cell as follows:

$$\frac{4\times\frac{4}{3}\pi r^3}{16\sqrt{2}r^3}=\frac{\pi}{3\sqrt{2}}=0.7405$$

The 74% packing efficiency is the maximum density possible in unit cells constructed of spheres of only one size. Most crystalline forms of metallic elements are hcp, fcc, or bcc (body-centred cubic). The coordination number of hcp and fcc is 12 and its atomic packing factor (APF) is the number mentioned above, 0.74. The APF of bcc is 0.68 for comparison.

Bravais Lattices

When the crystal systems are combined with the various possible lattice centerings, we arrive at the Bravais lattices. They describe the geometric arrangement of the lattice points, and thereby the translational symmetry of the crystal. In three dimensions, there are 14 unique Bravais lattices that are distinct from one another in the translational symmetry they contain. All crystalline materials recognized until now (not including quasicrystals) fit in one of these arrangements. The fourteen three-dimensional lattices, classified by crystal system. The Bravais lattices are sometimes referred to as *space lattices*.

The crystal structure consists of the same group of atoms, the *basis*, positioned around each and every lattice point. This group of atoms therefore repeats indefinitely in three dimensions according to the arrangement of one of the 14 Bravais lattices. The characteristic rotation and mirror symmetries of the group of atoms, or unit cell, is described by its crystallographic point group.

Point Groups

The crystallographic point group or *crystal class* is the mathematical group comprising the symmetry operations that leave at least one point unmoved and that leave the appearance of the crystal structure unchanged. These symmetry operations include

- *Reflection*, which reflects the structure across a *reflection plane*
- *Rotation*, which rotates the structure a specified portion of a circle about a *rotation axis*
- *Inversion*, which changes the sign of the coordinate of each point with respect to a *centre of symmetry* or *inversion point*
- *Improper rotation*, which consists of a rotation about an axis followed by an inversion.

Rotation axes (proper and improper), reflection planes, and centres of symmetry are collectively called *symmetry elements*. There are 32 possible crystal classes. Each one can be classified into one of the seven crystal systems.

Space Groups

The space group of the crystal structure is composed of the translational symmetry operations in addition to the operations of the point group. These include:

- Pure *translations*, which move a point along a vector
- *Screw axes*, which rotate a point around an axis while translating parallel to the axis
- *Glide planes*, which reflect a point through a plane while translating it parallel to the plane.

There are 230 distinct space groups.

Grain Boundaries

Grain boundaries are interfaces where crystals of different orientations meet. A grain boundary is a single-phase interface, with crystals on each side of the boundary being identical except in orientation. The term "crystallite boundary" is sometimes, though rarely, used. Grain boundary areas contain those atoms that have been perturbed from their original lattice sites, dislocations, and impurities that have migrated to the lower energy grain boundary. Treating a grain boundary geometrically as an interface of a single crystal cut into two parts, one of which is rotated, we see that there are five variables required to define a grain boundary.

The first two numbers come from the unit vector that specifies a rotation axis. The third number designates the angle of rotation of the grain. The final two numbers specify the plane of the grain boundary (or a unit vector that is normal to this plane).

Grain boundaries disrupt the motion of dislocations through a material, so reducing crystallite size is a common way to improve strength, as described by the Hall-Petch relationship. Since grain boundaries are defects in the crystal structure they tend to decrease the electrical and thermal conductivity of the material. The high interfacial energy and relatively weak bonding in most grain boundaries often makes them preferred sites for the onset of corrosion and for the precipitation of new phases from the solid. They are also important to many of the mechanisms of creep.

Grain boundaries are in general only a few nanometers wide. In common materials, crystallites are large enough that grain boundaries account for a small fraction of the material. However, very small grain sizes are achievable. In nanocrystalline solids, grain boundaries become a significant volume fraction of the material, with profound effects on such properties as diffusion and plasticity. In the limit of small crystallites, as the volume fraction of grain boundaries approaches 100%, the material ceases to have any crystalline character, and thus becomes an amorphous solid.

Defects and Impurities

Real crystals feature defects or irregularities in the ideal arrangements and it is these defects that critically determine many of the electrical and mechanical properties of real materials. When one atom substitutes for one of the principal atomic components within the crystal structure, alteration in the electrical and thermal properties of the material may ensue.

Impurities may also manifest as spin impurities in certain materials. Research on magnetic impurities demonstrates that substantial alteration of certain properties such as specific heat may be affected by small concentrations of an impurity, as for example impurities in semiconducting ferromagnetic alloys may lead to different properties as first predicted in the late 1960s. Dislocations in the crystal lattice allow shear at lower stress than that needed for a perfect crystal structure.

Prediction of Structure

The difficulty of predicting stable crystal structures based on the knowledge of only the chemical composition has long been a stumbling block on the way to fully computational materials design. Now, with more powerful algorithms and high-performance computing, structures of medium complexity can be predicted using such approaches as evolutionary algorithms, random sampling, or metadynamics.

The crystal structures of simple ionic solids (e.g., NaCl or table salt) have long been rationalized in terms of Pauling's rules, first set out in 1929 by Linus Pauling, referred to by many since as the "father of the chemical bond". Pauling also considered the nature of the interatomic forces in metals, and concluded that about half of the five d-orbitals in the transition metals are involved in bonding, with the remaining nonbonding d-orbitals being responsible for the magnetic properties. He, therefore, was able to correlate the number of d-orbitals in bond formation with the bond length as well as many of the physical properties of the substance. He subsequently introduced the metallic orbital, an extra orbital necessary to permit uninhibited resonance of valence bonds among various electronic structures.

In the resonating valence bond theory, the factors that determine the choice of one from among alternative crystal structures of a metal or intermetallic compound revolve around the energy of resonance of bonds among interatomic positions. It is clear that some modes of resonance would make larger contributions (be more mechanically stable than others), and that in particular a simple ratio of number of bonds to number of positions would be exceptional. The resulting principle is that a special stability is associated with the simplest ratios or "bond numbers": 1/2, 1/3, 2/3, 1/4, 3/4, etc. The choice of structure and the value of the axial ratio (which determines the relative bond lengths) are thus a result of the effort of an atom to use its valency in the formation of stable bonds with simple fractional bond numbers.

After postulating a direct correlation between electron concentration and crystal structure in beta-phase alloys, Hume-Rothery analyzed the trends in melting points, compressibilities and bond lengths as a function of group number in the periodic table in order to establish a system of valencies of the transition elements in the metallic state. This treatment thus emphasized the increasing bond strength as a function of group number. The operation of directional forces were emphasized in one article on the relation between bond hybrids and the metallic structures. The resulting correlation between electronic and crystalline structures is summarized by a single parameter, the weight of the d-electrons per hybridized metallic orbital. The “d-weight” calculates out to 0.5, 0.7 and 0.9 for the fcc, hcp and bcc structures respectively. The relationship between d-electrons and crystal structure thus becomes apparent.

Polymorphism

Figure 7: Quartz is one of the several thermodynamically stable crystalline forms of silica, SiO_2. The most important forms of silica include: a-quartz, b-quartz, tridymite, cristobalite, coesite, and stishovite.

Polymorphism refers to the ability of a solid to exist in more than one crystalline form or structure. According to Gibbs’ rules of phase equilibria, these unique crystalline phases will be dependent on intensive variables such as pressure and temperature. Polymorphism can potentially be found in many crystalline materials including polymers, minerals, and metals, and is related to allotropy, which refers to elemental solids. The complete morphology of a material is described by polymorphism and other variables such as crystal habit, amorphous fraction or crystallographic defects. Polymorphs have different stabilities and may spontaneously convert from a metastable form (or thermodynamically unstable form) to the stable form at a particular temperature. They also exhibit different melting points, solubilities, and X-ray diffraction patterns.

One good example of this is the quartz form of silicon dioxide, or SiO_2. In the vast majority of silicates, the Si atom shows tetrahedral coordination by 4 oxygens. All but one of the crystalline forms involve tetrahedral SiO_4 units linked together by shared vertices in different arrangements. In different minerals the tetrahedra show different degrees of networking and polymerization. For example, they occur singly, joined together in pairs, in larger finite clusters including rings, in chains, double chains, sheets, and three-dimensional frameworks. The minerals are classified into groups based on these structures. In each of its 7 thermodynamically stable crystalline forms or polymorphs of crystalline quartz, only 2 out of 4 of each the edges of the SiO_4 tetrahedra are shared with others, yielding the net chemical formula for silica: SiO_2.

Another example is elemental tin (Sn), which is malleable near ambient temperatures but is brittle when cooled. This change in mechanical properties due to existence of its two major allotropes, α- and β-tin. The two allotropes that are encountered at normal pressure and temperature, α-tin and β-tin, are more commonly known as *gray tin* and *white tin* respectively. Two more allotropes, γ and σ, exist at temperatures above 161 °C and pressures above several GPa. White tin is metallic, and is the stable crystalline form at or above room temperature. Below 13.2 °C, tin exists in the gray form, which has a diamond cubic crystal structure, similar to diamond, silicon or germanium. Gray tin has no metallic properties at all, is a dull-gray powdery material, and has few uses, other than a few specialized semiconductor applications. Although the α-β transformation temperature of tin is nominally 13.2 °C, impurities (e.g. Al, Zn, etc.) lower the transition temperature well below 0 °C, and upon addition of Sb or Bi the transformation may not occur at all.

Physical Properties

Twenty of the 32 crystal classes are so-called piezoelectric, and crystals belonging to one of these classes (point groups) display piezoelectricity. All piezoelectric classes lack a centre of symmetry. Any material develops a dielectric polarization when an electric field is applied, but a substance that has such a natural charge separation even in the absence of a field is called a polar material. Whether or not a material is polar is determined solely by its crystal structure. Only 10 of the 32 point groups are polar. All polar crystals are pyroelectric, so the 10 polar crystal classes are sometimes referred to as the pyroelectric classes.

There are a few crystal structures, notably the perovskite structure, which exhibit ferroelectric behaviour. This is analogous to ferromagnetism, in that, in the absence of an electric field during production, the ferroelectric crystal does not exhibit a polarization. Upon the application of an electric field of sufficient magnitude, the crystal becomes permanently polarized.

This polarization can be reversed by a sufficiently large countercharge, in the same way that a ferromagnet can be reversed. However, it is important to note that, although they are called ferroelectrics, the effect is due to the crystal structure (not the presence of a ferrous metal).

Crystal Habit

Crystal habit is an overall description of the visible external shape of a mineral. This description can apply to an individual crystal or an assembly of crystals or aggregates.

In mineralogy, shape and size give rise to descriptive terms applied to the typical appearance, or habit of crystals. Each crystal can be described by how well it is formed, ranging from euhedral (perfect to near-perfect), to subhedral (moderately formed), and anhedral (poorly formed to no discernable habit seen).

The many terms used by mineralogists to describe crystal habits are useful in communicating what specimens of a particular mineral often look like.

Figure 8: Goethite Replacing Pyrite Cubes

Recognizing numerous habits helps a mineralogist to identify a large number of minerals. Some habits are distinctive of certain minerals, although most minerals exhibit many differing habits (the development of a particular habit is determined by the details of the conditions during the mineral formation/crystal growth). Crystal habit may mislead the inexperienced as a mineral's internal crystal system can be hidden or disguised. Factors influencing a crystal's habit include: a combination of two or more crystal forms; trace impurities present during growth; crystal twinning and growth conditions (i.e., heat, pressure, space); and specific growth tendencies like growth striations. Minerals belonging to the same crystal system do not necessarily exhibit the same habit. Some habits of a mineral are unique to its variety and locality: For example, while most sapphires form elongate barrel-shaped crystals, those found in Montana form stout *tabular* crystals. Ordinarily, the latter habit is seen only in ruby. Sapphire and ruby are both varieties of the same mineral; corundum.

Some minerals may replace other existing minerals while preserving the original's habit: this process is called pseudomorphous replacement. A classic example is tiger's eye quartz, crocidolite asbestos replaced by silica. While quartz typically forms *prismatic* (elongate, prism-like) crystals, in tiger's eye the original *fibrous* habit of crocidolite is preserved. The names of crystal habits are derived from: Predominant crystal faces (prism-prismatic, pyramid-pyramidal and pinacoid-platy). Crystal forms (cubic, octahedral, dodecahedral). Aggregation of crystals or aggregates (fibrous, botryoidal, radiating, massive). Crystal appearance (foliated/lamellar (layered), dendritic, bladed, acicular, lenticular, tabular (tablet shaped)).

Crystal Twinning

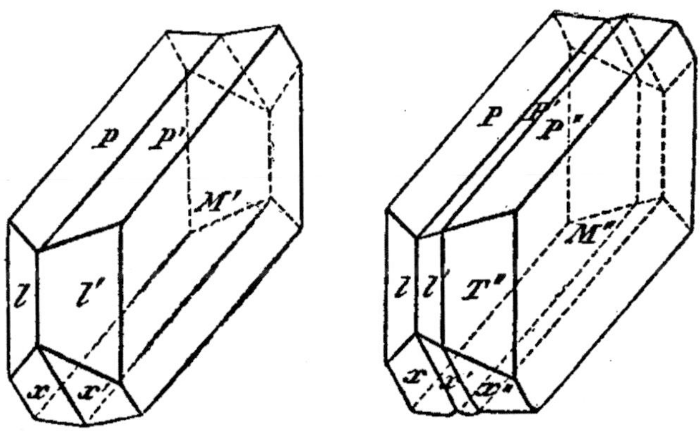

Figure 9: Diagram of twinned crystals of Albite. On the more perfect cleavage, which is parallel to the basal plane (P), is a system of fine striations, parallel to the second cleavage (M).

Crystal twinning occurs when two separate crystals share some of the same crystal lattice points in a symmetrical manner. The result is an intergrowth of two separate crystals in a variety of specific configurations. A twin boundary or composition surface separates the two crystals. Crystallographers classify twinned crystals by a number of twin laws. These twin laws are specific to the crystal system. The type of twinning can be a diagnostic tool in mineral identification.

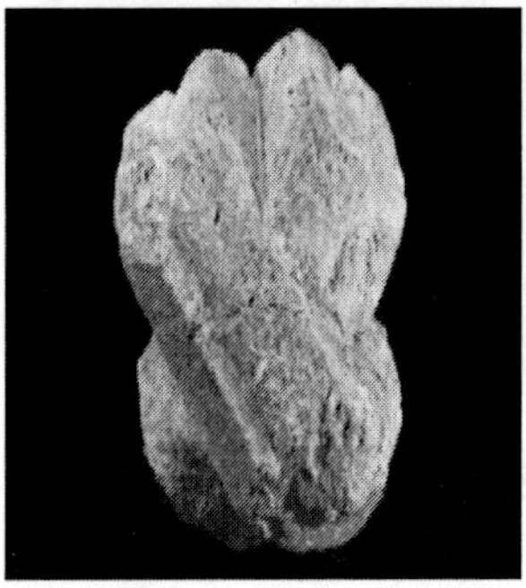

Figure 10: Penetration Twinning in Orthoclase, Carlsbad law

Simple twinned crystals may be *contact twins* or *penetration twins*. Contact twins share a single composition surface often appearing as mirror images across the boundary. Plagioclase, quartz, gypsum, and spinel often exhibit contact twinning. In penetration twins the individual crystals have the appearance of *passing through* each other in a symmetrical manner. Orthoclase, staurolite, pyrite, and fluorite often show penetration twinning.

If several twin crystal parts are aligned by the same twin law they are referred to as multiple or repeated twins. If these multiple twins are aligned in parallel they are called *polysynthetic twins*. When the multiple twins are not parallel they are *cyclic twins*.

Albite, calcite, and pyrite often show polysynthetic twinning. Closely spaced polysynthetic twinning is often observed as striations or fine parallel lines on the crystal face. Rutile, aragonite, cerussite, and chrysoberyl often exhibit cyclic twinning, typically in a radiating pattern.

There are three modes of formation of twinned crystals. *Growth twins* are the result of an interruption or change in the lattice during formation or growth due to a possible deformation from a larger substituting ion. *Annealing* or *Transformation twins* are the result of a change in crystal system during cooling as one *form* becomes unstable and the crystal structure must re-organize or *transform* into another more stable form. *Deformation* or *gliding twins* are the result of stress on the crystal after the crystal has formed. Deformation twinning is a common result of regional metamorphism.

Figure 11: Twinned Pyrite Crystal Group

Crystals that grow adjacent to each other may be aligned to resemble twinning. This *parallel growth* simply reduces system energy and is not twinning.

Twin Boundaries

Twin boundaries occur when two crystals of the same type intergrow, so that only a slight misorientation exists between them. It is a highly symmetrical interface, often with one crystal the mirror image of the other; also, atoms are shared by the two crystals at regular intervals. This is also a much lower-energy interface than the grain boundaries that form when crystals of arbitrary orientation grow together.

Twin boundaries are partly responsible for shock hardening and for many of the changes that occur in cold work of metals with limited slip systems or at very low temperatures. They also occur due to martensitic transformations: the motion of twin boundaries is responsible for the pseudoelastic and shape-memory behaviour of nitinol, and their presence is partly responsible for the hardness due to quenching of steel.

Deformation Twins

Of the three common crystalline structures BCC, FCC, and HCP, the HCP structure is the most likely to form deformation twins when strained, mainly due to the lack of slip systems in this structure.

Cleavage (crystal)

Cleavage, in mineralogy, is the tendency of crystalline materials to split along definite crystallographic structural planes. These planes of relative weakness are a result of the regular locations of atoms and ions in the crystal, which create smooth repeating surfaces that are visible both in the microscope and to the naked eye.

Types of cleavage

Cleavage forms parallel to crystallographic planes:

- Basal or pinacoidal cleavage occurs parallel to the base of a crystal. This orientation is given by the {001} plane in the crystal lattice, and is the same as the {0001} plane in Bravais-Miller indices, which are often used for rhombohedral and hexagonal crystals. Basal cleavage is exhibited by the mica group and by graphite.
- Cubic cleavage occurs on the {001} planes, parallel to the faces of a cube for a crystal with cubic symmetry. This is the source of the cubic shape seen in crystals of ground table salt, the mineral halite. The mineral galena also typically exhibits perfect cubic cleavage.
- Octahedral cleavage occurs on the {111} crystal planes, forming octahedra shapes for a crystal with cubic symmetry. Diamond and fluorite exhibit perfect octahedral cleavage. Octahedral cleavage is seen in common semiconductors. For lower-symmetry crystals, there will be a smaller number of {111} planes.
- Dodecahedral cleavage occurs on the {110} crystal planes forming dodecahedra for a crystal with cubic symmetry. For lower-symmetry crystals, there will be a smaller number of {110} planes.
- Rhombohedral cleavage occur parallel to the {1011} faces of a rhombohedron. Calcite and other carbonate minerals exhibit perfect rhombohedral cleavage.
- Prismatic cleavage is cleavage parallel to a vertical prism {110}. Cerussite, tremolite and spodumene exhibit prismatic cleavage.

Parting

Crystal parting occurs when minerals break along planes of structural weakness due to external stress or along twin composition planes. Parting breaks are very similar in appearance to cleavage, but only occur due to stress. Examples include magnetite which shows octahedral parting, the rhombohedral parting of corundum and basal parting in pyroxenes.

Uses

Cleavage is a traditional physical property used in mineral identification both in hand specimen and microscopic examination of rock and mineral studies. As an example, the angles between the prismatic cleavage planes for the pyroxenes (88–92°) and the amphiboles (56–124°) are diagnostic.

Crystal cleavage is of technical importance in the electronics industry and in the cutting of gemstones.

Precious stones are generally cleaved by impact as in diamond cutting.

Synthetic single crystals of semiconductor materials are generally sold as thin wafers which are much easier to cleave. Simply pressing a silicon wafer against a soft surface and scratching its edge with a diamond scribe is usually enough to cause cleavage; however, when dicing a wafer to form chips, a procedure of scoring and breaking is often followed for greater control. Elemental semiconductors (Si, Ge, and diamond) are diamond cubic, a space group for which octahedral cleavage is observed. This means that some orientations of wafer allow near-perfect rectangles to be cleaved. Most other commercial semiconductors (GaAs, InSb, etc.) can be made in the related zinc blende structure, with similar cleavage planes.

Lustre (mineralogy)

Lustre (or luster) is a description of the way light interacts with the surface of a crystal, rock, or mineral. The word lustre traces its origins back to the Latin word *lux*, meaning "light", and generally implies radiance, gloss, or brilliance.

A range of terms are used to describe lustre. Many terms refer to materials having similar lustres, such as *earthy*, *metallic*, *greasy*, and *silky*. Similarly, the term *vitreous* (derived from the Latin for glass, *vitrum*) refers to a glassy lustre. A list of these terms is given below. Lustre varies over a wide continuum, and so there are no rigid boundaries between the different terms. (For this reason, different sources can often describe the same mineral differently. This ambiguity is further complicated by the fact that lustre can vary widely within a particular mineral species.)

The terms are frequently combined to describe intermediate types of lustre (for example, a "vitreous greasy" lustre).

Some minerals exhibit unusual optical phenomena, such as asterism (the display of a star-shaped luminous area) or chatoyancy (the display of luminous bands, which appear to move as the specimen is rotated). A list of such phenomena is given below.

Common Terms

Adamantine Lustre

Adamantine minerals possess a superlative lustre, which is most notably seen in diamond. Such minerals are transparent or translucent, and have a high refractive index (of 1.9 or more). Minerals with a true adamantine lustre are uncommon, with examples being cerussite and zircon. Minerals with a lesser (but still relatively high) degree of lustre are referred to as subadamantine, with some examples being garnet and corundum.

Dull Lustre

Figure 12 : Kaolinite

Dull (or earthy) minerals exhibit little to no lustre, due to coarse granulations which scatter light in all directions, approximating a Lambertian reflector. An example is kaolinite. A distinction is sometimes drawn between dull minerals and earthy minerals, with the latter being coarser, and having even less lustre.

Greasy Lustre

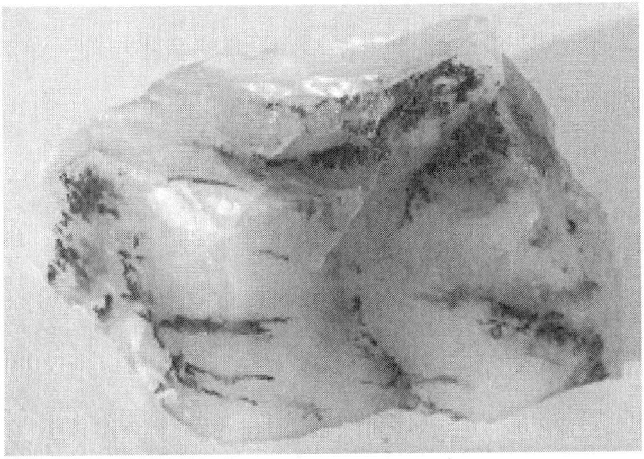

Figure 13 : Moss Opal

Greasy minerals resemble fat or grease. A greasy lustre often occurs in minerals containing a great abundance of microscopic inclusions, with examples including opal and cordierite. Many minerals with a greasy lustre are also greasy to the touch.

Metallic Luster

Metallic (or splendent) minerals have the lustre of polished metal (and with ideal surfaces will act like a mirror). Examples include galena, pyrite and magnetite.

Pearly Luster

Figure 14 : Muscovite

Pearly minerals consist of thin transparent co-planar sheets. Light reflecting from these layers give them a lustre reminiscent of pearls. Such minerals possess perfect cleavage, with examples including muscovite and stilbite.

Resinous Lustre

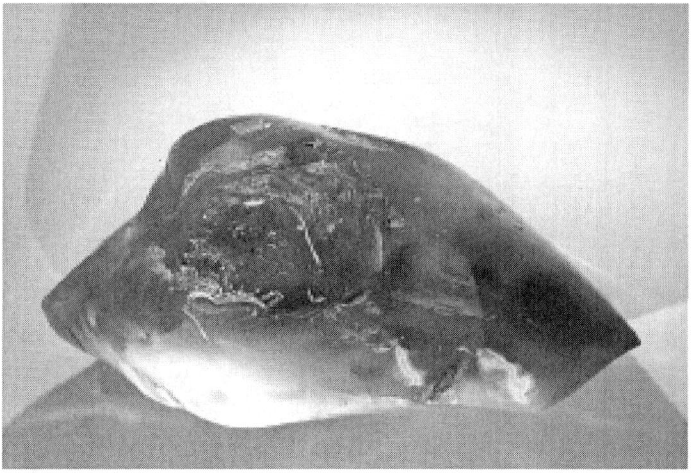

Figure 15 : Amber

Resinous minerals have the appearance of resin, chewing gum or (smooth surfaced) plastic. A principal example is amber, which is a form of fossilized resin.

Silky Lustre

Figure 16 : Satin Spar Variety of Gypsum

Silky minerals have a parallel arrangement of extremely fine fibres, giving them a lustre reminiscent of silk. Examples include asbestos, ulexite and the satin spar variety of gypsum. A fibrous lustre is similar, but has a coarser texture.

Submetallic Lustre

Figure 17 : Sphalerite

Submetallic minerals have similar lustre to metal, but are duller and less reflective. A submetallic lustre often occurs in near-opaque minerals with very high refractive indices, such as sphalerite, cinnabar and cuprite.

Vitreous Lustre

Figure 18 : Quartz

Vitreous minerals have the lustre of glass. (The term is derived from the Latin for glass, *vitrum*.) This type of lustre is one of the most commonly seen, and occurs in transparent or translucent minerals with relatively low refractive indices. Common examples include calcite, quartz, topaz, beryl, tourmaline and fluorite, among others.

Waxy Lustre

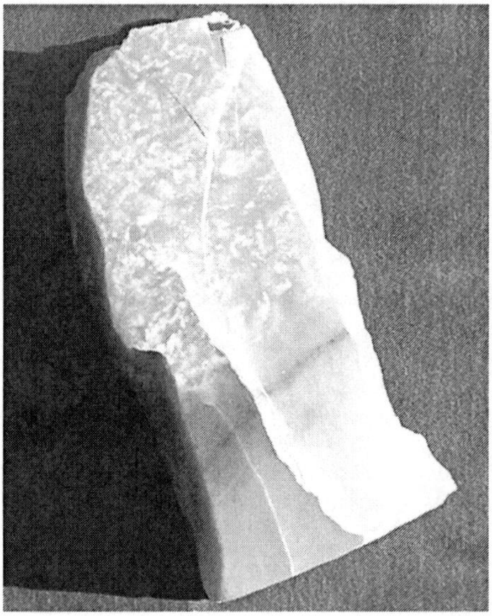

Figure 19 : Jade

Waxy minerals have a lustre resembling wax. Examples include jade and chalcedony.

Optical Phenomena

Asterism

Figure 20 : Sapphire Cabochon

Asterism is the display of a star-shaped luminous area. It is seen in some sapphires and rubies, where it is caused by impurities of rutile. It can also occur in garnet, diopside and spinel.

Aventurescence

Figure 21 : Aventurine

Aventurescence (or aventurization) is a reflectance effect like that of glitter. It arises from minute, preferentially oriented mineral platelets within the material. These platelets are so numerous that they also influence the material's body colour. In aventurine quartz chrome-bearing fuchsite makes for a green stone, and various iron oxides make for a red stone.

Chatoyancy

Chatoyant minerals display luminous bands, which appear to move as the specimen is rotated. Such minerals are composed of parallel fibres (or contain fibrous voids or inclusions), which reflect light into a direction perpendicular to their orientation, thus forming narrow bands of light. The most famous examples are tiger's eye and cymophane, but the effect may also occur in other minerals such as aquamarine, moonstone and tourmaline.

Colour Change

Colour change is most commonly found in Alexandrite, a variety of chrysoberyl gemstones. Other gems also occur in colour-change varieties, including (but not limited to) sapphire, garnet, spinel. Alexandrite displays a colour change dependent upon light, along with strong pleochroism. The gem results from small scale replacement of aluminium by chromium oxide, which is responsible for alexandrite's characteristic green to red colour change. Alexandrite from the Ural Mountains in Russia is green by daylight and red by incandescent light. Other varieties of alexandrite may be yellowish or pink in daylight and a columbine or raspberry red by incandescent light. The optimum or "ideal" colour change would be fine emerald green to fine purplish red, but this is exceedingly rare.

Schiller

Schiller, from German for "twinkle", is a term used to describe the metallic iridescence originating from below the surface of a stone, that occurs when light is reflected between layers of minerals. It is seen in moonstone and labradorite and is very similar to adularescence and aventurescence.

Streak (Mineralogy)

The streak (also called "powder colour") of a mineral is the colour of the powder produced when it is dragged across an unweathered surface. Unlike the apparent colour of a mineral, which for most minerals can vary considerably, the trail of finely ground powder generally has a more consistent characteristic colour, and is thus an important diagnostic tool in mineral identification. If no streak seems to be made, the mineral's streak is said to be white or colourless. Streak is particularly important as a diagnostic for opaque and coloured materials. It is less useful for silicate minerals, most of which have a white streak and are too hard to powder easily.

The apparent colour can vary widely because of trace impurities or a disturbed macroscopic crystal structure. Small amounts of an impurity that strongly absorbs a particular wavelength can radically change the wavelengths of light that are reflected by the specimen, and thus change the apparent colour. However, when the specimen is dragged to produce a streak, it is broken into randomly oriented microscopic crystals, and small impurities do not greatly affect the absorption of light.

The surface across which the mineral is dragged is called a "streak plate," and is generally made of unleased porcelain tile. In the absence of a streak plate, the unleased underside of a porcelain bowl or vase or the back of a glazed tile will work. Sometimes a streak is more easily or accurately described by comparing it with the "streak" made by another streak plate. Because the trail left behind results from the mineral being crushed into powder, a streak can only be made of minerals softer than the streak plate, around 7 on the Mohs scale of mineral hardness. In case of harder minerals, the colour of the powder can be determined by filing or crushing with a hammer a small sample, which is then usually rubbed on a streak plate. Most minerals that are harder have an unhelpful white streak. Some minerals leave a streak similar to their natural colour, such as cinnabar and lazurite. Other minerals leave surprising colours, such as fluorite, which always has a white streak, although it can appear in purple, blue, yellow, or green crystals. Hematite, which is black in appearance, leaves a red streak which accounts for its name, which comes from the Greek word "haima," meaning "blood." Galena, which can be similar in appearance to hematite, is easily distinguished by its gray streak.

Mohs Scale of Mineral Hardness

The Mohs scale of mineral hardness characterizes the scratch resistance of various minerals through the ability of a harder material to scratch a softer material. It was created in 1812 by the German geologist and mineralogist Friedrich Mohs and is one of several definitions of hardness in materials science. The method of comparing hardness by seeing which minerals can scratch others, however, is of great antiquity, having first been mentioned by Theophrastus in his treatise *On Stones*, *circa* 300 BC, followed by Pliny the Elder in his *Naturalis Historia*, *circa* 77 AD.

Minerals

The Mohs scale of mineral hardness is based on the ability of one natural sample of matter to scratch another. The samples of matter used by Mohs are all minerals. Minerals are pure substances found in nature. Rocks are made up of one or more minerals. As the hardest known naturally occurring substance when the scale was designed, diamonds are at the top of the scale. The hardness of a material is measured against the scale by finding the hardest material that the given material can scratch, and/or the softest material that can scratch the given material. For example, if some material is scratched by apatite but not by fluorite, its hardness on the Mohs scale would fall between 4 and 5.

The Mohs scale is a purely ordinal scale. For example, corundum (9) is twice as hard as topaz (8), but diamond (10) is four times as hard as corundum. The table below shows comparison with absolute hardness measured by a sclerometer, with pictorial examples. On the

Mohs scale, graphite (a principal constituent of pencil "lead") has a hardness of 1.5; a fingernail, 2.2–2.5; a copper penny, 3.2–3.5; a pocketknife 5.1; a knife blade, 5.5; window glass plate, 5.5; and a steel file, 6.5. A streak plate (unleased porcelain) has a hardness of 7.0. Using these ordinary materials of known hardness can be a simple way to approximate the position of a mineral on the scale.

Intermediate Hardness

The table below incorporates additional substances that may fall between levels:

Hardness	*Substance or Mineral*
0.2–0.3	caesium, rubidium
0.5–0.6	lithium, sodium, potassium
1	talc
1.5	gallium, strontium, indium, tin, barium, thallium, lead, graphite
2	hexagonal boron nitride, calcium, selenium, cadmium, sulfur, tellurium, bismuth
2.5 to 3	magnesium, gold, silver, aluminium, zinc, lanthanum, cerium, Jet_(lignite) (lignite)
3	calcite, copper, arsenic, antimony, thorium, dentin
4	fluorite, iron, nickel
4 to 4.5	platinum, steel
5	apatite, cobalt, zirconium, palladium, tooth enamel, obsidian (volcanic glass)
5.5	beryllium, molybdenum, hafnium
6	orthoclase, titanium, manganese, germanium, niobium, rhodium, uranium
6 to 7	glass, fused quartz, iron pyrite, silicon, ruthenium, iridium, tantalum, opal
7	quartz, vanadium, osmium, rhenium
7.5 to 8	hardened steel, tungsten, emerald, spinel
8	topaz, cubic zirconia
8.5	chrysoberyl, chromium, silicon nitride
9-9.5	corundum, silicon carbide (carborundum), tungsten carbide, titanium carbide, stishovite
9.5–10	rhenium diboride, tantalum carbide, titanium diboride, boron nitride, boron.
10	diamond
>10	nanocrystalline diamond (hyperdiamond, ultrahard fullerite)

Hardness (Vickers)

Comparison between Hardness (Mohs) and Hardness (Vickers):

Mineralname	*Hardness (Mohs)*	*Hardness (Vickers)kg/mm²*
Graphite	1- 2	VHN_{10}=7-11
Tin	1½-2	VHN_{10}=7-9
Bismuth	2- 2½	VHN_{100}=16- 18
Gold	2½- 3	VHN_{10}=30- 34
Silver	2½- 3	VHN_{100}=61- 65
Chalcocite	2½- 3	VHN_{100}=84- 87
Copper	2½- 3	VHN_{100}=77- 99
Galena	2½	VHN_{100}=79- 104
Sphalerite	3½- 4	VHN_{100}=208- 224
Heazlewoodite	4	VHN_{100}=230- 254
Carrollite	4½- 5½	VHN_{100}=507- 586
Goethite	5- 5½	VHN_{100}=667
Hematite	5- 6	VHN_{100}=1,000- 1,100
Chromite	5½	VHN_{100}=1,278- 1,456
Anatase	5½- 6	VHN_{100}=616- 698
Rutile	6- 6½	VHN_{100}=894- 974
Pyrite	6- 6½	VHN_{100}=1,505- 1,520
Bowieite	7	VHN_{100}=858- 1,288
Euclase	7½	VHN_{100}=1,310
Chromium	9	VHN_{100}=1,875- 2,000

Crystallography

Crystallography is the experimental science of the arrangement of atoms in solids. The word "crystallography" derives from the Greek words *crystallon* = cold drop/frozen drop, with its meaning extending to all solids with some degree of transparency, and *grapho* = write.

Before the development of X-ray diffraction crystallography, the study of crystals was based on their geometry. This involves measuring the angles of crystal faces relative to theoretical reference axes (crystallographic axes), and establishing the symmetry of the crystal in question.

The former is carried out using a goniometer. The position in 3D space of each crystal face is plotted on a stereographic net, e.g. Wulff net or Lambert net. In fact, the pole to each face is plotted on the net. Each point is labelled with its Miller index. The final plot allows the symmetry of the crystal to be established.

Crystallographic methods now depend on the analysis of the diffraction patterns of a sample targeted by a beam of some type. Although X-rays are most commonly used, the beam is not always electromagnetic radiation.

For some purposes electrons or neutrons are used. This is facilitated by the wave properties of the particles. Crystallographers often explicitly state the type of illumination used when referring to a method, as with the terms *X-ray diffraction, neutron diffraction* and *electron diffraction.*

These three types of radiation interact with the specimen in different ways. X-rays interact with the spatial distribution of the valence electrons, while electrons are charged particles and therefore feel the total charge distribution of both the atomic nuclei and the surrounding electrons.

Neutrons are scattered by the atomic nuclei through the strong nuclear forces, but in addition, the magnetic moment of neutrons is non-zero. They are therefore also scattered by magnetic fields. When neutrons are scattered from hydrogen-containing materials, they produce diffraction patterns with high noise levels. However, the material can sometimes be treated to substitute hydrogen for deuterium. Because of these different forms of interaction, the three types of radiation are suitable for different crystallographic studies.

Theory

Generally, an image of a small object is made using a lens to focus the illuminating radiation, as is done with the rays of the visible spectrum in light microscopy. However, the wavelength of visible light (about 4000 to 7000 angstroms) is three orders of magnitude longer than the length of typical atomic bonds and atoms themselves (about 1 to 2 angstroms).

Therefore, obtaining information about the spatial arrangement of atoms requires the use of radiation with shorter wavelengths, such as X-rays or neutrons beam. Employing shorter wavelengths implied abandoning microscopy and true imaging, however, because there exists no material from which a lens capable of focusing this type of radiation can be created. (That said, scientists have had some success focusing X-rays with microscopic Fresnel zone plates made from gold, and by critical-angle reflection inside long tapered capillaries.) Diffracted x-ray or neutrons beams cannot be focused to produce images, so the sample structure must be reconstructed from the diffraction pattern.

Sharp features in the diffraction pattern arise from periodic, repeating structure in the sample, which are often very strong due to coherent reflection of many photons from many regularly spaced instances of similar structure, while non-periodic components of the structure result in diffuse (and usually weak) diffraction features. Said more simply, areas with a higher density and repetition of atom order tend to reflect more light toward one point in space when compared to those areas with fewer atoms and less repetition.

Because of their highly ordered and repetitive structure, crystals give diffraction patterns of sharp Bragg reflection spots, and are ideal for analysing the structure of solids.

Notation

- Coordinates in *square brackets* such as [100] denote a direction vector (in real space).
- Coordinates in *angle brackets* or *chevrons* such as <100> denote a *family* of directions which are related by symmetry operations. In the cubic crystal system for example, <100> would mean [100], [010], [001] or the negative of any of those directions.
- Miller indices in *parentheses* such as (100) denote a plane of the crystal structure, and regular repetitions of that plane with a particular spacing. In the cubic system, the normal to the (hkl) plane is the direction [hkl], but in lower-symmetry cases, the normal to (hkl) is not parallel to [hkl].
- Indices in *curly brackets* or *braces* such as {100} denote a family of planes and their normals which are equivalent in cubic materials due to symmetry operations, much the way angle brackets denote a family of directions. In non-cubic materials, <hkl> is not necessarily perpendicular to {hkl}.

Technique

Some materials studied using crystallography, proteins for example, do not occur naturally as crystals. Typically, such molecules are placed in solution and allowed to crystallize over days, weeks, or months through vapour diffusion. A drop of solution containing the molecule, buffer, and precipitants is sealed in a container with a reservoir containing a hygroscopic solution. Water in the drop diffuses to the reservoir, slowly increasing the concentration and allowing a crystal to form. If the concentration were to rise more quickly, the molecule would simply precipitate out of solution, resulting in disorderly granules rather than an orderly and hence usable crystal.

Once a crystal is obtained, data can be collected using a beam of radiation. Although many universities that engage in crystallographic research have their own X-ray producing equipment, synchrotrons are often used as X-ray sources, because of the purer and more complete patterns such sources can generate.

Synchrotron sources also have a much higher intensity of X-ray beams, so data collection takes a fraction of the time normally necessary at weaker sources. Complementary neutron crystallography techniques are used to enhance hydrogen atoms positions. Such techniques are available in Neutron facilities.

Producing an image from a diffraction pattern requires sophisticated mathematics and often an iterative process of modelling and refinement.

In this process, the mathematically predicted diffraction patterns of an hypothesized or "model" structure are compared to the actual pattern generated by the crystalline sample. Ideally, researchers make several initial guesses, which through refinement all converge on the same answer. Models are refined until their predicted patterns match to as great a degree as can be achieved without radical revision of the model. This is a painstaking process, made much easier today by computers.

The mathematical methods for the analysis of diffraction data only apply to *patterns,* which in turn result only when waves diffract from orderly arrays. Hence crystallography applies for the most part only to crystals, or to molecules which can be coaxed to crystallize for the sake of measurement. In spite of this, a certain amount of molecular information can be deduced from the patterns that are generated by fibres and powders, which while not as perfect as a solid crystal, may exhibit a degree of order. This level of order can be sufficient to deduce the structure of simple molecules, or to determine the coarse features of more complicated molecules. For example, the double-helical structure of DNA was deduced from an X-ray diffraction pattern that had been generated by a fibrous sample.

Crystallography in Materials Engineering

Crystallography is a tool that is often employed by materials scientists. In single crystals, the effects of the crystalline arrangement of atoms is often easy to see macroscopically, because the natural shapes of crystals reflect the atomic structure.

In addition, physical properties are often controlled by crystalline defects. The understanding of crystal structures is an important prerequisite for understanding crystallographic defects. Mostly, materials do not occur in a single crystalline, but poly-crystalline form, such that the powder diffraction method plays a most important role in structural determination.

A number of other physical properties are linked to crystallography. For example, the minerals in clay form small, flat, platelike structures. Clay can be easily deformed because the platelike particles can slip along each other in the plane of the plates, yet remain strongly connected in the direction perpendicular to the plates. Such mechanisms can be studied by crystallographic texture measurements.

In another example, iron transforms from a body-centred cubic (bcc) structure to a face-centred cubic (fcc) structure called austenite when it is heated. The fcc structure is a close-packed structure, and the bcc structure is not, which explains why the volume of the iron decreases when this transformation occurs.

Crystallography is useful in phase identification. When performing any process on a material, it may be desired to find out what compounds and what phases are present in the material. Each phase has a characteristic arrangement of atoms. Techniques like X-ray or neutron diffraction can be used to identify which patterns are present in the material, and thus which compounds are present.

Crystallography covers the enumeration of the symmetry patterns which can be formed by atoms in a crystal and for this reason has a relation to group theory and geometry.

Biology

X-ray crystallography is the primary method for determining the molecular conformations of biological macromolecules, particularly protein and nucleic acids such as DNA and RNA. In fact, the double-helical structure of DNA was deduced from crystallographic data. The first crystal structure of a macromolecule was solved in 1958 A three-dimensional model of the myoglobin molecule obtained by X-ray analysis. The Protein Data Bank (PDB) is a freely accessible repository for the structures of proteins and other biological macromolecules. Computer programmes like RasMol or Pymol can be used to visualize biological molecular structures.

Neutron crystallography is often used to help refine structures obtained by x-ray methods or to solve a specific bond; the methods are often viewed as complementary, as x-rays are sensitive to electron positions and scatter most strongly off heavy atoms, while neutrons are sensitive to nucleus positions and scatter strongly off many light isotopes, including hydrogen and deuterium. Electron crystallography has been used to determine some protein structures, most notably membrane proteins and viral capsids.

Chapter 3

Optical Mineralogy

Optical mineralogy is the study of minerals and rocks by measuring their optical properties. Most commonly, rock and mineral samples are prepared as thin sections or grain mounts for study in the laboratory with a petrographic microscope. Optical mineralogy is used to identify the mineralogical composition of geological materials in order to help reveal their origin and evolution.

Some of the properties and techniques used include:

Refractive Index

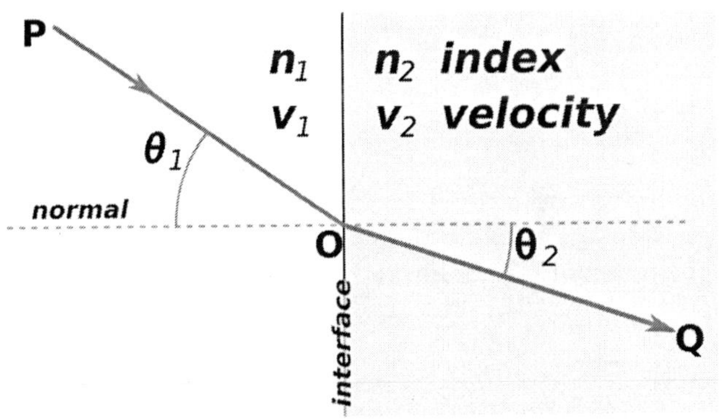

Figure 1 : Refraction of Light at the Interface Between two Media.

In optics the refractive index or index of refraction of a substance or medium is a measure of the speed of light in that medium. It is expressed as a ratio of the speed of light in vacuum relative to that in the considered medium. This can be written mathematically as:

n = speed of light in a vacuum/speed of light in medium.

For example, the refractive index of water is 1.33, meaning that light travels 1.33 times as fast in vacuum as it does in water.

As light moves from a medium, such as air, water, or glass, into another it may change its propagation direction in proportion to the change in refractive index. This refraction is governed by Snell's law, and is illustrated in the figure to the right. Refractive index of materials varies with the wavelength of light. This is called dispersion and results in a slightly different refractive index for each colour.

The wavelength λ of light in a material is determined by the refractive index according to $\lambda = \lambda_0/n$, where λ_0 is the wavelength of the light in vacuum. Brewster's angle, the critical angle for total internal reflection, and the reflectivity of a surface is also affected by the refractive index. These material parameters can be calculated using the Fresnel equations.

The concept of refractive index is widely used within the full electromagnetic spectrum, from x-rays to radio waves. It can also be used with wave phenomena other than light, e.g. sound. In this case the speed of sound is used instead of that of light and a reference medium other than vacuum must be chosen.

Definitions

The refractive index, n, of a medium is defined as the ratio of the speed, c, of a wave phenomenon such as light or sound in a reference medium to the phase speed, v_p, of the wave in the medium in question:

$$n = \frac{c}{v_p}.$$

It is most commonly used in the context of light with vacuum as a reference medium, although historically other reference media (e.g. air at a standardized pressure and temperature) have been common. It is usually given the symbol n. In the case of light, it equals

$$n = \sqrt{\grave{o}_r \mu_r},$$

where $\grave{o}_r$ is the material's relative permittivity, and μ_r is its relative permeability. For most naturally occurring materials, μ_r is very close to 1 at optical frequencies, therefore n is approximately $\sqrt{\grave{o}_r}$. Contrary to a widespread misconception, the real part of a complex n may be less than one, depending upon the material and wavelength. This has practical technical applications, such as effective mirrors for X-rays based on total external reflection. The phase speed is defined as the rate at which the crests of the waveform propagate; that is, the rate at which the phase of the waveform is moving. The *group speed* is the rate at which the *envelope* of the waveform is propagating; that is, the rate of variation of the amplitude of the waveform. Provided the waveform is not distorted significantly during propagation, it is the group speed that represents the rate at which information (and energy) may be transmitted by the wave (for example, the speed at which a pulse of light travels down an optical fibre). For the analytic properties constraining the unequal phase and group speeds in dispersive media, refer to dispersion (optics).

Another common definition of the refractive index comes from the refraction of a light ray entering a medium. The refractive index is the ratio of the sines of the angles of incidence θ_1 and refraction θ_2 as light passes into the medium or mathematically

$$n = \frac{\sin\theta_1}{\sin\theta_2}.$$

The angles are measured to the normal of the surface. This definition is based on Snell's law and is equivalent to the definition above if the light enters from the reference medium (normally vacuum).

A complex refractive index is often used to take absorption into account. This is further discussed in the Dispersion and absorption section below.A closely related quantity is *refractivity*, which in atmospheric applications is denoted N and defined as $N = 10^6(n- 1)$. The 10^6 factor is used because for air, n deviates from unity at most a few parts per thousand.

Typical Values

Selected refractive indices at λ = 589nm (more extensive list here).

Material		***n***
	Gases at 0 °C and 1 atm	
Air		1.000293
Helium		1.000036
Hydrogen		1.000132
Carbon dioxide		1.00045
	Liquids at 20 °C	
Water		1.333
Ethanol		1.36

Methanol	1.329
Benzene	1.501
Glycerine	1.473
Solids	
Ice	1.309
Fused silica	1.46
PMMA	1.49
Crown glass (typical)	1.52
Flint glass (typical)	1.62
Diamond	2.42

For visible light most transparent media have refractive indices between 1 and 2. A few examples are given in the table to the right. These values are measured at the yellow doublet sodium D-line, with a wavelength of 589 nanometres, as is conventionally done. For infrared light refractive indices can be considerably higher. Germanium is transparent in this region and has a refractive index of about 4, making it an important material for infrared optics.

Refractive Index Below 1

A widespread misconception is that since, according to the theory of relativity, nothing can travel faster than the speed of light in vacuum, the refractive index cannot be lower than 1. This is erroneous since the refractive index measures the phase velocity of light, which does not carry energy or information, the two things limited in propagation speed.

The phase velocity is the speed at which the crests of the wave move and can be faster than the speed of light in vacuum, and thereby give a refractive index below 1.

This can occur close to resonance frequencies, in plasmas, and for x-rays. In the x-ray regime the refractive indices are lower than but very close to 1 (exceptions close to some resonance frequencies). As an example, water has a refractive index of $1–2.6\times10^{-7}$ at a photon energy of 30 keV (0.04 nm wavelength).

Negative Refractive Index

Recent research has also demonstrated the existence of the negative refractive index, which can occur if permittivity and permeability have simultaneous negative values. This can be achieved with periodically constructed metamaterials. The resulting negative refraction (i.e., a reversal of Snell's law) offers the possibility of the superlens and other exotic phenomena.

Microscopic Explanation

At the microscale, an electromagnetic wave's phase speed is slowed in a material because the electric field creates a disturbance in the charges of each atom (primarily the electrons) proportional to the electric susceptibility of the medium. (Similarly, the magnetic field creates a disturbance proportional to the magnetic susceptibility.) As the electromagnetic fields oscillate in the wave, the charges in the material will be "shaken" back and forth at the same frequency. The charges thus radiate their own electromagnetic wave that is at the same frequency, but usually with a phase delay, as the charges may move out of phase with the force driving them.

The light wave travelling in the medium is the macroscopic superposition (sum) of all such contributions in the material: The original wave plus the waves radiated by all the moving charges.

This wave is typically a wave with the same frequency but shorter wavelength than the original, leading to a slowing of the wave's phase speed. Most of the radiation from oscillating material charges will modify the incoming wave, changing its velocity. However, some net energy will be radiated in other directions or even at other frequencies.

Depending on the relative phase of the original driving wave and the waves radiated by the charge motion, there are several possibilities:

- If the electrons emit a light wave which is 90° out of phase with the light wave shaking them, it will cause the total light wave to travel more slowly. This is the normal refraction of transparent materials like glass or water, and corresponds to a refractive index which is real and greater than 1.
- If the electrons emit a light wave which is 270° out of phase with the light wave shaking them, it will cause the total light wave to travel more quickly. This is called "anomalous refraction", and is observed close to absorption lines, with X-rays, and in some microwave systems. It corresponds to a refractive index less than 1. If the response is sufficiently strong and out-of-phase.
- If the electrons emit a light wave which is 180° out of phase with the light wave shaking them, it will destructively interfere

with the original light to reduce the total light intensity. This is light absorption in opaque materials and corresponds to an imaginary refractive index.

- If the electrons emit a light wave which is in phase with the light wave shaking them, it will amplify the light wave. This is rare, but occurs in lasers due to stimulated emission. It corresponds to an imaginary index of refraction, with the opposite sign as absorption.

For most materials at visible-light frequencies, the phase is somewhere between 90° and 180°, corresponding to a combination of both refraction and absorption.

Dispersion

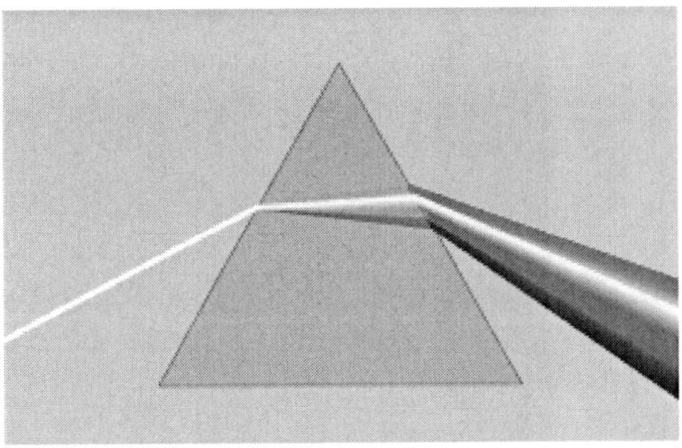

Figure 2 : In a prism dispersion causes different colors to refract at different angles, splitting white light into a rainbow of colors.

The refractive index of materials varies with the wavelength (and frequency) of light. This is called dispersion and causes prisms to divide white light into its constituent spectral colours, and explains how rainbows are formed. As the refractive index varies with wavelength, according to Snell's law, so will the refraction angle as light goes from one material to another. This makes different colours go in different directions. Dispersion also causes the focal length of lenses to be wavelength dependent. This is a type of chromatic aberration, which often needs to be corrected for in imaging systems.

In regions of the spectrum where the material does not absorb, the refractive index tends to decrease with increasing wavelength, and thus increase with frequency. This is called normal dispersion, in contrast to anomalous dispersion, where the refractive index increases with wavelength. For visible light normal dispersion means that the refractive index is higher for blue light than for red.

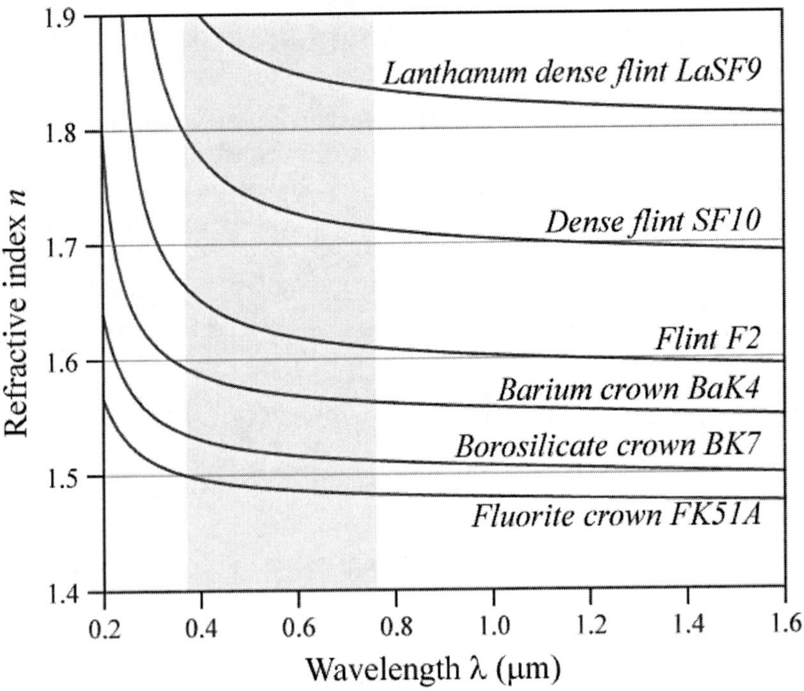

Figure 3: The variation of refractive index with wavelength for various glasses.

For optics in the visual range the amount of dispersion of a lens material is often quantified by the Abbe number $V = \frac{n_{yellow} - 1}{n_{blue} - n_{red}}$. For a more accurate description of the wavelength dependence of the refractive index the Sellmeier equation can be used. It is an empirical formula that works well in describing dispersion. *Sellmeier coefficients* are often quoted instead of the refractive index in tables.

Because of dispersion, it is usually important to specify the vacuum wavelength at which a refractive index is measured. Typically, this is done at various well-defined spectral emission lines; for example, n_D is the refractive index at the Fraunhofer "D" line, the centre of the yellow sodium double emission at 589.29 nm wavelength.

Complex Index of Refraction and Absorption

When light passes through a medium, some part of it will always be absorbed. This can be conveniently taken into account by defining a complex index of refraction,

$$\tilde{n} = n + i\kappa.$$

Here, the real part of the refractive index n indicates the phase speed, while the imaginary part κ indicates the amount of absorption loss when the electromagnetic wave propagates through the material.

That κ corresponds to absorption can be seen by inserting this refractive index into the expression for electric field of a plane electromagnetic wave travelling in the z-direction. We can do this by relating the wave number to the refractive index through $k = \frac{2\pi n}{\lambda_0}$, with λ_0 being the vacuum wavelength. With complex wave number $\tilde{k}$ and refractive index $n + i\kappa$ this can be inserted into the plane wave expression as:

$$\mathbf{E}(z,t) = \mathrm{Re}(\mathbf{E}_0 e^{i(\tilde{k}z-\omega t)}) = \mathrm{Re}(\mathbf{E}_0 e^{i(2\pi(n+i\kappa)z/\lambda_0-\omega t)}) = e^{-2\pi\kappa z/\lambda_0}\mathrm{Re}(\mathbf{E}_0 e^{i(kz-\omega t)}).$$

Here we see that κ gives an exponential decay, as expected from the Beer–Lambert law.

κ is often called the extinction coefficient in physics although this has a different definition within chemistry. Both n and κ are dependent on the frequency.

In most circumstances $\kappa > 0$ (light is absorbed) or $\kappa = 0$ (light travels forever without loss). In special situations, especially in the gain medium of lasers, it is also possible that $\kappa < 0$, corresponding to an amplification of the light.

An alternative convention uses $\tilde{n} = n - i\kappa$ instead of $\tilde{n} = n + i\kappa$, but where $\kappa > 0$ still corresponds to loss. Therefore these two conventions are inconsistent and should not be confused. The difference is related to defining sinusoidal time dependence as Re($e^{-i\omega t}$) versus Re($e^{+i\omega t}$).

Dielectric loss and non-zero DC conductivity in materials cause absorption. Good dielectric materials such as glass have extremely low DC conductivity, and at low frequencies the dielectric loss is also negligible, resulting in almost no absorption (κ H" 0). However, at higher frequencies (such as visible light), dielectric loss may increase absorption significantly, reducing the material's transparency to these frequencies.

The real and imaginary parts of the complex refractive index are related through the Kramers–Kronig relations. For example, one can determine a material's full complex refractive index as a function of wavelength from an absorption spectrum of the material.

For X-ray and extreme ultraviolet radiation the complex refractive index deviates only slightly from unity and usually has a real part smaller than 1. It is therefore normally written as $\tilde{n} = 1 - \delta + i\beta$ (or $\tilde{n} = 1 - \delta - i\beta$).

Relations to Other Quantities

Refraction

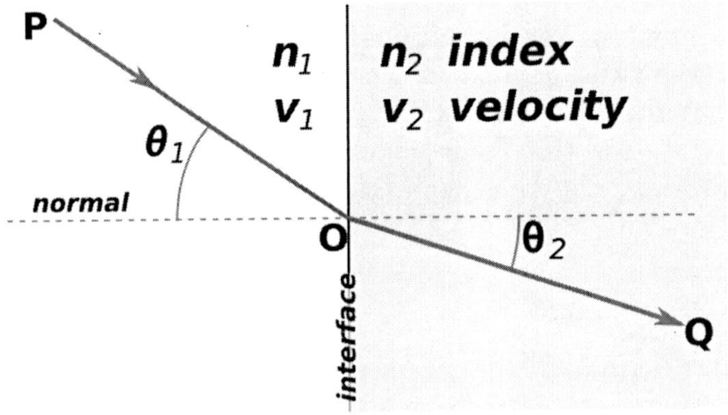

Refraction of light at the interface between two media of different refractive indices, with $n_2 > n_1$. Since the phase velocity is lower in the second medium ($v_2 < v_1$), the angle of refraction θ_2 is less than the angle of incidence θ_1; that is, the ray in the higher-index medium is closer to the normal.

When light moves from one medium to another as in the figure to the right, it changes direction, i.e. it is refracted. If it goes from a medium with refractive index n_1 to one with refractive index n_2, with an incidence angle to the surface normal of θ_1, the transmission angle θ_2 can be calculated from Snell's law:

$$n_1 \sin \theta_1 = n_2 \sin \theta_2.$$

If there is no angle θ_2 fulfilling Snell's law, i.e.

$$\frac{n_1}{n_2} \sin \theta_1 > 1$$

the light cannot be transmitted and will instead undergo total internal reflection.

Reflectivity

Apart from the transmitted light there is also a reflected part. The reflection angle is equal to the incidence angle, and the amount of light that is reflected is determined by the reflectivity of the surface.

The reflectivity can be calculated from the refractive index and the incidence angle with the Fresnel equations, which for normal incidence reduces to

$$R = \left(\frac{n_1 - n_2}{n_1 + n_2}\right)^2.$$

For common glass in air, $n_1 = 1$ and $n_2 = 1.5$, and thus about 4% of the incident power is reflected. At other incidence angles the reflectivity will also depend on the polarization of the incoming light. At a certain angle called Brewster's angle, p-polarized light (light with the electric field in the plane of incidence) will be totally transmitted. Brewster's angle can be calculated from the two refractive indices of the interface as

$$\theta_B = \arctan\left(\frac{n_2}{n_1}\right).$$

Lenses

The focal length of a lens is determined by its refractive index n and the radii of curvature R_1 and R_2 of its surfaces. The power of a thin lens in air is given by the Lensmaker's formula:

$$\frac{1}{f} = (n-1)\left(\frac{1}{R_1} - \frac{1}{R_2}\right).$$

Dielectric Constant

The frequency dependent dielectric constant is simply the square of the (complex) refractive index in a non-magnetic medium (one with a relative permeability of unity). The refractive index is used for optics in Fresnel equations and Snell's law; while the dielectric constant is used in Maxwell's equations and electronics.

Where $\tilde{\delta}$ is the complex dielectric constant with real and imaginary parts $\in_1$ and $\in_2$, and n and κ are the real and imaginary parts of the refractive index, all functions of frequency:

$$\tilde{\delta} = \grave{o}_1 + i\grave{o}_2 = (n + i\kappa)^2.$$

Conversion between refractive index and dielectric constant is done by:

$$\grave{o}_1 = n^2 - \kappa^2$$

$$\grave{o}_2 = 2n\kappa$$

$$n = \sqrt{\frac{\sqrt{\grave{o}_1^2 + \grave{o}_2^2} + \grave{o}_1}{2}}$$

$$\kappa = \sqrt{\frac{\sqrt{\grave{o}_1^2 + \grave{o}_2^2} - \grave{o}_1}{2}}.$$

Density

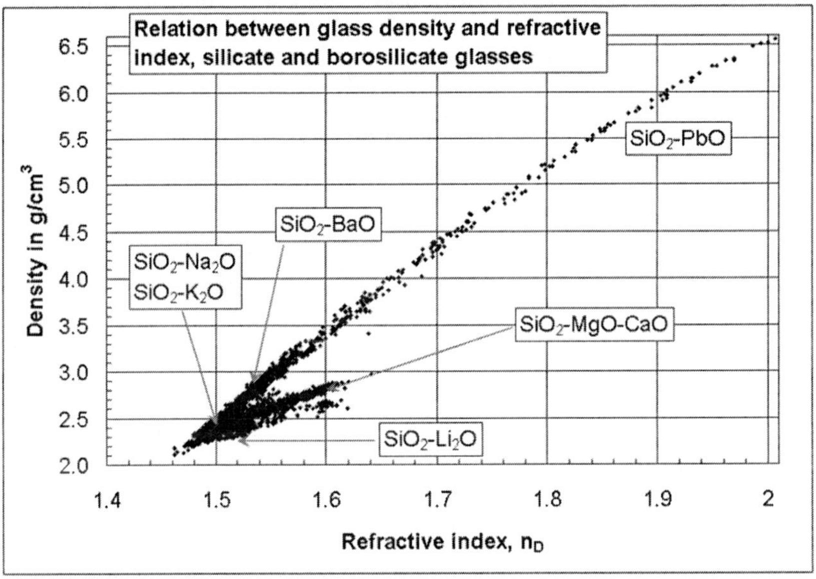

Figure 4: Relation between the refractive index and the density of silicate and borosilicate glasses.

In general, the refractive index of a glass increases with its density. However, there does not exist an overall linear relation between the refractive index and the density for all silicate and borosilicate glasses.

A relatively high refractive index and low density can be obtained with glasses containing light metal oxides such as Li_2O and MgO, while the opposite trend is observed with glasses containing PbO and BaO as seen in the diagram at the right.

Group Index

Sometimes, a "group speed refractive index", usually called the *group index* is defined:

$$n_g = \frac{c}{v_g}$$

where v_g is the group velocity. This value should not be confused with n, which is always defined with respect to the phase velocity. When the dispersion is small, the group velocity can be linked to the phase velocity by the relation

$$v_g = v - \lambda \frac{dv}{d\lambda}.$$

In this case the group index can thus be written in terms of the wavelength dependence of the refractive index as

$$n_g = \frac{n}{1 + \frac{\lambda}{n}\frac{dn}{d\lambda}},$$

where λ is the wavelength in the medium.

When the refractive index of a medium is known as a function of the vacuum wavelength (instead of the wavelength in the medium), the corresponding expressions for the group velocity and index are (for all values of dispersion)

$$v_g = c\left(n - \lambda_0 \frac{dn}{d\lambda_0}\right)^{-1}.$$

$$n_g = n - \lambda_0 \frac{dn}{d\lambda_0},$$

where λ_0 is the wavelength in vacuum.

Other Relations

As shown in the Fizeau experiment, when light is transmitted through a moving medium, its speed relative to a stationary observer is:

$$V = \frac{c}{n} + v\left(1 - \frac{1}{n^2}\right)$$

The refractive index of a substance can be related to its polarizability with the Lorentz–Lorenz equation or to the molar refractivities of its constituents by the Gladstone–Dale relation.

Anisotropy

The refractive index of certain media may be different depending on the polarization and direction of propagation of the light through the medium. This is known as birefringence or anisotropy and is described by the field of crystal optics. In the most general case, the *dielectric constant* is a rank-2 tensor (a 3 by 3 matrix), which cannot simply be described by refractive indices except for polarizations along principal axes.

In magneto-optic (gyro-magnetic) and optically active materials, the principal axes are complex (corresponding to elliptical polarizations), and the dielectric tensor is complex-Hermitian (for lossless media); such materials break time-reversal symmetry and are used e.g. to construct Faraday isolators.

In crystalline calcium carbonate (calcite), the birefringent (uniaxial) optics depend on directional differences in the structure. The index of refraction also depends on composition, and can be calculated using the Gladstone–Dale relation.

Nonlinearity

The strong electric field of high intensity light (such as output of a laser) may cause a medium's refractive index to vary as the light passes through it, giving rise to nonlinear optics. If the index varies quadratically with the field (linearly with the intensity), it is called the optical Kerr effect and causes phenomena such as self-focusing and self-phase modulation. If the index varies linearly with the field (which is only possible in materials that do not possess inversion symmetry), it is known as the Pockels effect.

Inhomogeneity

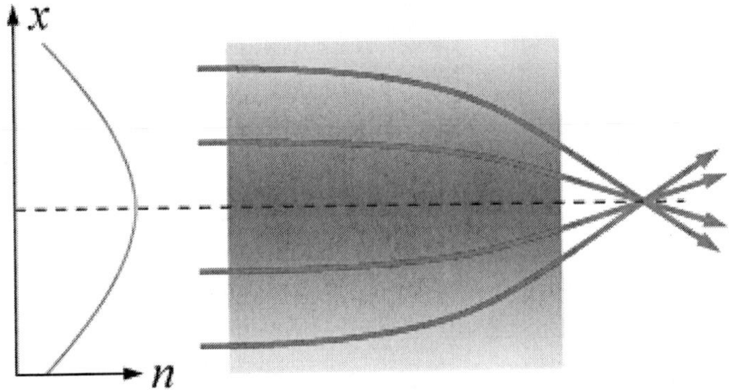

Figure 5: A gradient-index lens with a parabolic variation of refractive index (n) with radial distance (x). The lens focuses light in the same way as a conventional lens.

If the refractive index of a medium is not constant, but varies gradually with position, the material is known as a gradient-index medium and is described by gradient index optics. Light travelling through such a medium can be bent or focused, and this effect can be exploited to produce lenses, some optical fibres and other devices. Some common mirages are caused by a spatially-varying refractive index of air.

Refractive Index Measurement

Homogeneous Media

The refractive index of liquids or solids can be measured with refractometers. They typically measure some angle of refraction or the critical angle for total internal reflection. The first laboratory refractometers sold commercially where developed by Ernst Abbe in the late 19th century.

The same principles are still used today. In this instrument a thin layer of the liquid to be measured is placed between two prisms. Light is shone through the liquid at incidence angles all the way up to 90°, i.e. light rays parallel to the surface. The second prism should have an index of refraction higher than that of the liquid, so that light only enters the prism at angles smaller than the critical angle for total reflection. This angle can then be measured either by looking through a telescope, or with a digital photodetector placed in the focal plane of a lens. The refractive index n of the liquid can then be calculated from the maximum transmission angle θ as $n = n_G \sin \theta$, where n_G is the refractive index of the prism.

This type of devices are commonly used in chemical laboratories for identification of substances and for quality control. Handheld variants are used in agriculture by e.g. wine makers to determine sugar content in grape juice, and inline process refractometers are used in e.g. chemical and pharmaceutical industry for process control.

In gemology a different type of refractometer is used to measure index of refraction and birefringence of gemstones. The gem is placed on a high refractive index prism and illuminated from below. A high refractive index contact liquid is used to achieve optical contact between the gem and the prism. At small incidence angles most of the light will be transmitted into the gem, but at high angles total internal reflection will occur in the prism. The critical angle is normally measured by looking through a telescope.

Refractive Index Variations

To measure the spatial variation of refractive index in a sample phase-contrast imaging methods are used. These methods measure the variations in phase of the light wave exiting the sample. The phase is proportional to the optical path length the light ray has traversed, and thus gives a measure of the integral of the refractive index along the ray path. The phase cannot be measured directly at optical or higher frequencies, and therefore needs to be converted into intensity by interference with a reference beam. In the visual spectrum this is done using Zernike phase-contrast microscopy, differential interference contrast microscopy (DIC) or interferometry.

Zernike phase-contrast microscopy introduces a phase shift to the low spatial frequency components of the image with a phase-shifting annulus in the Fourier plane of the sample, so that higher frequency parts of the image can interfere with the low frequency reference beam.

In DIC the illumination is split up into two beams that are given different polarizations, are phase shifted differently, and are shifted transversely with slightly different amounts. After the specimen the two parts are made to interfere giving an image of the derivative of the optical path length in the direction of the difference in transverse shift. In interferometry the illumination is split up into two beams by a partially reflective mirror. One of the beams is let through the sample before they are combined to interfere and give a direct image of the phase shifts. If the optical path length variations are more than a wavelength the image will contain fringes.

There exist several x-ray phase-contrast imaging techniques to determine 2D or 3D spatial distribution of refractive index of samples in the x-ray regime.

Abraham˘Minkowski Controversy

In 1908, Hermann Minkowski calculated the momentum of a refracted ray, p, where E is energy of the photon, c is the speed of light in vacuum and n is the refractive index of the medium as follows:

$$p = \frac{nE}{c}.$$

In 1909, Max Abraham proposed the following formula for this calculation:

$$p = \frac{E}{nc}.$$

A 2010 study suggested that *both* equations are correct, with the Abraham version being the kinetic momentum and the Minkowski version being the canonical momentum, and claims to explain the contradicting experimental results using this interpretation.

Applications

The refractive index of a material is the most important property of any optical system that uses refraction. It is used to calculate the focusing power of lenses, and the dispersive power of prisms. It can also be used as a useful tool to differentiate between different types of gemstone, due to the unique chatoyance each individual stone displays.

Since refractive index is a fundamental physical property of a substance, it is often used to identify a particular substance, confirm its purity, or measure its concentration. Refractive index is used to measure solids (glasses and gemstones), liquids, and gases. Most commonly it is used to measure the concentration of a solute in an aqueous solution. A refractometer is the instrument used to measure refractive index. For a solution of sugar, the refractive index can be used to determine the sugar content.

In GPS, the index of refraction is utilized in ray-tracing to account for the radio propagation delay due to the Earth's electrically neutral atmosphere. It is also used in Satellite link design for the Computation of radiowave attenuation in the atmosphere

Birefringence

Birefringence, or double refraction, is the decomposition of a ray of light into two rays when it passes through certain anisotropic materials, such as crystals of calcite or boron nitride. The effect was first described by the Danish scientist Rasmus Bartholin in 1669, who saw it in calcite. The effect is now known to also occur in certain plastics, magnetic materials, various noncrystalline materials, and liquid crystals.

The simplest instance of the effect arises in materials with uniaxial anisotropy. That is, the structure of the material is such that it has an axis of symmetry with no equivalent axis in the plane perpendicular to it. (Cubic crystals are thereby ruled out.) This axis is known as the optical axis of the material, and light with linear polarizations parallel and perpendicular to it has unequal indices of refraction, denoted n_e and n_o, respectively, where the suffixes stand for extraordinary and ordinary.

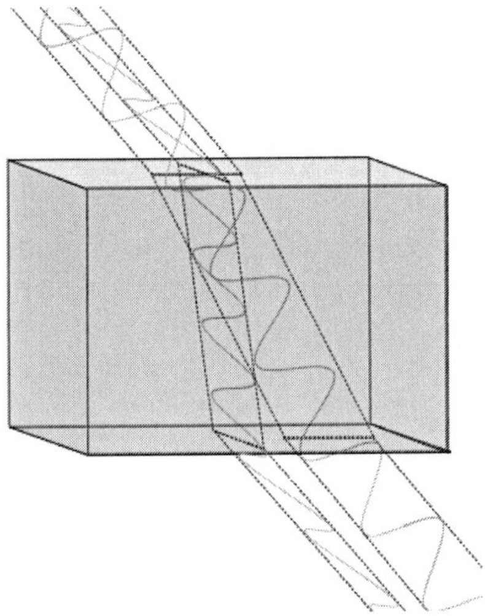

Figure 6: Displacement of light rays with perpendicular polarization through a birefringent material.

The names reflect the fact that, if unpolarized light enters the material at a nonzero acute angle to the optical axis, the component with polarization perpendicular to this axis will be refracted as per the standard law of refraction, while the complementary polarization component will refract at a nonstandard angle determined by the angle of entry and the difference between the indices of refraction,

$$\Delta n = n_e - n_o$$

known as the birefringence magnitude. The light will therefore split into two linearly polarized beams, known as ordinary and extraordinary. Exceptions arise when the light propagates either along or orthogonal to the optical axis. In the first case, both polarizations and rays are ordinary and are not split. In the second case also, there is no splitting of the light into two separate directions, but the ordinary and extraordinary components travel at different speeds, and the effect is used to interconvert between linear and circular or elliptical polarizations.

Double refraction also occurs in biaxially anisotropic materials, which are also known as trirefringent, but its description is then substantially more complex.

Creation

While birefringence is often found naturally (especially in crystals), there are several ways to create it in optically isotropic materials.

- Birefringence results when isotropic materials are deformed such that the isotropy is lost in one direction (i.e., stretched or bent). Example
- Applying an electric field can induce molecules to line up or behave asymmetrically, introducing anisotropy and resulting in birefringence.
- Applying a magnetic field can cause a material to be circularly birefringent, with different indices of refraction for oppositely-handed circular polarizations
- Self alignment of highly polar molecules such as lipids and some surfactants will generate highly birefringent thin films (see also Liquid crystal).

Examples of Uniaxial Birefringent Materials

Table 1 : Uniaxial materials, at 590 nm.

Material	n_o	n_e	Δn
beryl $Be_3Al_2(SiO_3)_6$	1.602	1.557	-0.045
calcite $CaCO_3$	1.658	1.486	-0.172
calomel Hg_2Cl_2	1.973	2.656	+0.683
ice H_2O	1.309	1.313	+0.004
lithium niobate $LiNbO_3$	2.272	2.187	-0.085
magnesium fluoride MgF_2	1.380	1.385	+0.006
quartz SiO_2	1.544	1.553	+0.009
ruby Al_2O_3	1.770	1.762	-0.008
rutile TiO_2	2.616	2.903	+0.287
peridot $(Mg, Fe)_2SiO_4$	1.690	1.654	-0.036
sapphire Al_2O_3	1.768	1.760	-0.008
sodium nitrate $NaNO_3$	1.587	1.336	-0.251
tourmaline (complex silicate)	1.669	1.638	-0.031
zircon, high $ZrSiO_4$	1.960	2.015	+0.055
zircon, low $ZrSiO_4$	1.920	1.967	+0.047

The best-studied uniaxial birefringent materials are crystalline, the refractive indices of several of which tabulated to the right (at wavelength ~ 590 nm). Silicon carbide, also known as Moissanite, is strongly birefringent.

Many plastics are birefringent, because their molecules are 'frozen' in a stretched conformation when the plastic is moulded or extruded. For example, cellophane is a cheap birefringent material, and Polaroid sheets are commonly used to examine for orientation in birefringent plastics like polystyrene and polycarbonate. Birefringent materials are used in many devices that manipulate the polarization of light, such as wave plates, polarizing prisms, and Lyot filters.

As stated above, birefringence can also arise in magnetic materials, but substantial variations in magnetic permeability of materials are rare at optical frequencies.

Birefringence can be observed in amyloid plaque deposits such as are found in the brains of Alzheimer's patients. Modified proteins such as immunoglobulin light chains abnormally accumulate between cells, forming fibrils. Multiple folds of these fibres line up and take on a beta-pleated sheet conformation. Congo red dye intercalates between the folds and, when observed under polarized light, causes birefringence.

Cotton (Gossypium hirsutum) fibre is birefringent because of high levels of cellulosic material in the fibre's secondary cell wall.

Slight imperfections in optical fibre can cause birefringence, which can cause distortion in fibre-optic communication; see polarization mode dispersion. The imperfections can be geometrically based, or a result of photoelastic effects from loading on the optical fibre.

Fast and Slow Rays

Table 2 : Effective refractive indices in negative uniaxial materials

Propagation direction	***Ordinary ray***		***Extraordinary ray***	
	Polarization	n_{eff}	***Polarization***	n_{eff}
z	xy-plane	n_o	n/a	n/a
xy-plane	xy-plane	n_o	z	n_e
xz-plane	y	n_o	xz-plane	$n_e < n < n_o$
other	analogous to xz-plane			

For a given propagation direction, in general there are two perpendicular polarizations for which the medium behaves as if it had a single effective refractive index. In a uniaxial material, rays with these polarizations are called the extraordinary and the ordinary ray (*e* and *o* rays), corresponding to the extraordinary and ordinary refractive indices. In a biaxial material, there are three refractive indices α, β, and γ, yet only two rays, which are called the fast and the slow ray. The slow ray is the ray that has the highest effective refractive index.

For a uniaxial material with the z axis defined to be the optical axis, the effective refractive indices are as in the table on the right. For rays propagating in the xz plane, the effective refractive index of the e polarization varies continuously between n_o and n_e, depending on the angle with the z axis. The effective refractive index can be constructed from the Index ellipsoid.

Biaxial Birefringence

Table 3: Biaxial materials, at 590 nm.

Material	n_α	n_β	n_γ
borax	1.447	1.469	1.472
epsom salt $MgSO_4 \cdot 7(H_2O)$	1.433	1.455	1.461
mica, biotite	1.595	1.640	1.640
mica, muscovite	1.563	1.596	1.601
olivine $(Mg, Fe)_2SiO_4$	1.640	1.660	1.680
perovskite $CaTiO_3$	2.300	2.340	2.380
topaz	1.618	1.620	1.627
ulexite	1.490	1.510	1.520

Biaxial birefringence, also known as trirefringence, describes an anisotropic material that has more than one axis of anisotropy. For such a material, the refractive index tensor n, will in general have three distinct eigenvalues that can be labeled n_α, n_β and n_γ.

Positive or Negative

Uniaxial birefringent materials are classified as positively (or negatively) birefringent when, for light (having parallel and perpendicular components) directed to the optic axis, the refractive index of light polarized parallel to the optic axis is greater (or smaller, respectively) than light polarized perpendicularly to the optic axis. In other words, the polarization of the slow (or fast) wave is parallel to the optical axis when the birefringence of the crystal is positive (or negative, respectively).

Biaxial crystals are defined as positively (or negatively) birefringent when the slow ray (or fast ray, respectively) bisects the acute angle formed by the optical axes.

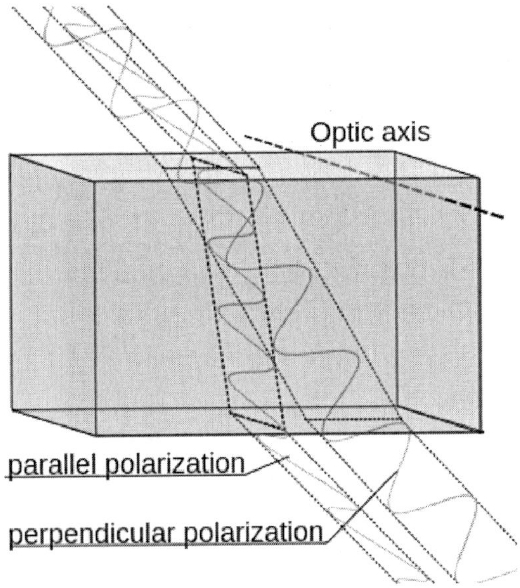

Figure 7 : Rays passing through a positively birefringent material. The incident light is having parallel and perpendicular polarisation components (unpolarised light). The optical axis is perpendicular to the direction of the perpendicular component of incident ray, so the ray polarized parallel to the optic axis has a greater refractive index than the ray polarized perpendicular to it.

In practice, when using an optical compensator that emits red light, a crystal with positive birefringence appears blue when its long dimension is parallel to the slow axis of the compensator. In contrast, a crystal with negative birefringence appears yellow when its long dimension is parallel to the slow axis of the compensator, and the slow ray of the compensator is oriented perpendicularly to the long axis of the crystal.

The reason for these phenomena is that the wavelength of emitted light is shifted higher in positively birefringent crystals, because the slow ray of the crystal is parallel to the slow axis of the compensator, while for negatively birefringent crystals the wavelength of emitted light is shifted lower, because the fast ray of the crystal is parallel to the slow axis of the compensator. The order of colours resulting from the use of compensators in a polarized light system differs from that of a typical spectrum, instead having an order including yellow-orange-red-violet-blue.

Measurement

Birefringence and related optical effects (such as optical rotation and linear or circular dichroism) can be measured by measuring the changes in the polarization of light passing through the material. These measurements are known as polarimetry.

Birefringence of lipid bilayers can be measured using dual polarisation interferometry. This provides a measure of the degree of order within these fluid layers and how this order is disrupted when the layer interacts with other biomolecules.

A common feature of optical microscopes is a pair of crossed polarizing filters. Between the crossed polarizers, a birefringent sample will appear bright against a dark (isotropic) background.

For a fixed composition such as calcium carbonate, a crystal such as calcite or its polymorphs, the index of refraction depends on the direction of light through the crystal structure. The refraction also depends on composition, and can be calculated using the Gladstone-Dale relation.

Applications

Birefringence is widely used in optical devices, such as liquid crystal displays, light modulators, colour filters, wave plates, optical axis gratings, etc. It also plays an important role in second harmonic generation and many other nonlinear processes.

Birefringent filters are also used as spatial low-pass filters in electronic cameras, where the thickness of the crystal is controlled to spread the image in one direction, thus increasing the spot-size. This is essential to the proper working of all television and electronic film cameras, to avoid spatial aliasing, the folding back of frequencies higher than can be sustained by the pixel matrix of the camera.

Medicine

Needle aspiration of fluid from a gouty joint will reveal negatively birefringent urate crystals. Calcium pyrophosphate crystals, in contrast, show weak positive birefringence. In practice, urate crystals appear yellow and calcium pyrophosphate crystals appear blue when their long axes are aligned parallel to that of a red compensator filter, or a crystal of known birefringence is added to the sample for comparison. In ophthalmology, scanning laser polarimetry utilises the birefringence of the retinal nerve fibre layer to indirectly quantify its thickness, which is of use in the assessment and monitoring of glaucoma.

Birefringence characteristics in sperm heads allow for the selection of spermatozoa for intracytoplasmic sperm injection. Likewise, *zona imaging* uses birefringence on oocytes to select the ones with highest chances of successful pregnancy. Birefringence of particles biopsied from pulmonary nodules indicates silicosis.

Elastic Birefringence

Another form of birefringence is observed in anisotropic elastic materials. In these materials, shear waves split according to similar principles as the light waves. The study of birefringent shear waves in the earth is a part of seismology. Birefringence is also used in optical mineralogy to determine the chemical composition, and history of minerals and rocks.

Stress Induced Birefringence

Isotropic solids do not exhibit birefringence. However, when they are under mechanical stress, birefringence results. The stress can be applied externally or is 'frozen' in after a birefringent plastic ware is cooled after it is manufactured using injection molding. When such a sample is placed between two crossed polarizers, colour patterns can be observed, because polarization of a light ray is rotated after passing through a birefingent material and the amount of rotation is dependent on wavelength. The experimental method called photoelasticity used for analysing stress distribution in solids is based on the same principle.

Theory

More generally, birefringence can be defined by considering a dielectric permittivity and a refractive index that are tensors. Consider a plane wave propagating in an anisotropic medium, with a relative permittivity tensor ε, where the refractive index n, is defined by $\mathrm{n}\cdot\mathrm{n} = \grave{o}$. If the wave has an electric vector of the form:

$$\mathbf{E} = \mathbf{E}_0 \exp i(\mathbf{k}\cdot\mathbf{r} - \omega t)$$

where r is the position vector and *t* is time, then the wave vector k and the angular frequency ω must satisfy Maxwell's equations in the medium, leading to the equations:

$$-\nabla \times \nabla \times \mathbf{E} = \frac{1}{c^2}\left(\grave{\mathbf{o}}\cdot\frac{\partial^2 \mathbf{E}}{\partial t^2}\right) \qquad \text{(a)}$$

$$\nabla\cdot(\varepsilon\cdot\mathbf{E}) = 0 \tag{b}$$

where c is the speed of light in a vacuum. Substituting eqn. 2 in eqns. a-b leads to the conditions:

$$|\mathbf{k}|^2\,\mathbf{E}_0 - (\mathbf{k}\cdot\mathbf{E}_0)\mathbf{k} = \frac{\omega^2}{c^2}(\varepsilon\cdot\mathbf{E}_0) \tag{a}$$

$$\mathbf{k}\cdot(\varepsilon\cdot\mathbf{E}_0) = 0 \tag{b}$$

For the matrix product ($\in$.E)often a separate name is used, the *dielectric displacement vector* D. So essentially birefringence concerns the general theory of linear relationships between these two vectors in anisotropic media.

To find the allowed values of k, E_0 can be eliminated from eq 4a. One way to do this is to write eqn 4a in Cartesian coordinates, where the x, y and z axes are chosen in the directions of the eigenvectors of ε, so that

$$\varepsilon = \begin{bmatrix} n_x^2 & 0 & 0 \\ 0 & n_y^2 & 0 \\ 0 & 0 & n_z^2 \end{bmatrix} \tag{c}$$

Hence eqn 4a becomes

$$\left(-k_y^2 - k_z^2 + \frac{\omega^2 n_x^2}{c^2}\right)E_x + k_x k_y E_y + k_x k_z E_z = 0 \tag{a}$$

$$k_x k_y E_x + \left(-k_x^2 - k_z^2 + \frac{\omega^2 n_y^2}{c^2}\right)E_y + k_y k_z E_z = 0 \tag{b}$$

$$k_x k_z E_x + k_y k_z E_y + \left(-k_x^2 - k_y^2 + \frac{\omega^2 n_z^2}{c^2}\right)E_z = 0 \tag{c}$$

where E_x, E_y, E_z, k_x, k_y and k_z are the components of E_0 and k. This is a set of linear equations in E_x, E_y, E_z, and they have a non-trivial solution if their determinant is zero:

$$\det\begin{bmatrix} \left(-k_y^2 - k_z^2 + \frac{\omega^2 n_x^2}{c^2}\right) & k_x k_y & k_x k_z \\ k_x k_y & \left(-k_x^2 - k_z^2 + \frac{\omega^2 n_y^2}{c^2}\right) & k_y k_z \\ k_x k_z & k_y k_z & \left(-k_x^2 - k_y^2 + \frac{\omega^2 n_z^2}{c^2}\right) \end{bmatrix} = 0$$

Multiplying out eqn, and rearranging the terms, we obtain

$$\frac{\omega^4}{c^4}-\frac{\omega^2}{c^2}\left(\frac{k_x^2+k_y^2}{n_z^2}+\frac{k_x^2+k_z^2}{n_y^2}+\frac{k_y^2+k_z^2}{n_x^2}\right)+\left(\frac{k_x^2}{n_y^2n_z^2}+\frac{k_y^2}{n_x^2n_z^2}+\frac{k_z^2}{n_x^2n_y^2}\right)(k_x^2+k_y^2+k_z^2)=0$$

In the case of a uniaxial material, where $n_x=n_y=n_o$ and $n_z=n_e$ say, eqn 7 can be factorised into

$$\left(\frac{k_x^2}{n_o^2}+\frac{k_y^2}{n_o^2}+\frac{k_z^2}{n_o^2}-\frac{\omega^2}{c^2}\right)\left(\frac{k_x^2}{n_e^2}+\frac{k_y^2}{n_e^2}+\frac{k_z^2}{n_o^2}-\frac{\omega^2}{c^2}\right)=0$$

Each of the factors in eqn 8 defines a surface in the space of vectors k — the surface of wave normals. The first factor defines a sphere and the second defines an ellipsoid. Therefore, for each direction of the wave normal, two wavevectors k are allowed. Values of k on the sphere correspond to the ordinary rays while values on the ellipsoid correspond to the extraordinary rays.

For a biaxial material, eqn cannot be factorized in the same way, and describes a more complicated pair of wave-normal surfaces.

Birefringence is often measured for rays propagating along one of the optical axes (or measured in a two-dimensional material). In this case, n has two eigenvalues that can be labeled n_1 and n_2. n can be diagonalized by:

$$\mathbf{n}=\mathbf{R}(\chi)\cdot\begin{bmatrix} n_1 & 0 \\ 0 & n_2 \end{bmatrix}\cdot\mathbf{R}(\chi)^T$$

where R(χ) is the rotation matrix through an angle χ. Rather than specifying the complete tensor n, we may now simply specify the *magnitude* of the birefringence Δn, and *extinction angle* χ, where $\Delta n = n_1 - n_2$.

Vacuum Birefringence

Due to quantum effects, the vacuum can behave like a birefringent medium. In the limit of weak fields and slowly varying fields, the Euler-Heisenberg Lagrangian becomes:

$$L=\frac{1}{2}\left(\mathbf{E}^2-\mathbf{B}^2\right)+\frac{2\alpha^2}{45m^4}\left[\left(\mathbf{E}^2-\mathbf{B}^2\right)^2+7\left(\mathbf{E}\cdot\mathbf{B}\right)^2\right]$$

where α is the fine sructure constant and m the electron mass. The effective electric permittivity tensor and magnetic permeability tensor are then given by :

$$\grave{o}_{ik} = \left[1 + \frac{8\alpha^2}{45m^4}\left(\mathbf{E}^2 - \mathbf{B}^2\right)\right]\delta_{ik} + \frac{28\alpha^2}{45m^4}B_iB_k$$

$$\mu_{ik} = \left[1 - \frac{8\alpha^2}{45m^4}\left(\mathbf{E}^2 - \mathbf{B}^2\right)\right]\delta_{ik} + \frac{28\alpha^2}{45m^4}E_iE_k$$

In the case of an plane wave propagating through a region of constant magnetic field, one finds that there are two eigenmodes. One has the electric component of the plane wave perpendicular to the plane spanned by the wavevector and the magnetic field, the other has the magnetic component perpendicular to this plane. The former has an index of refraction of $1 + \frac{8\alpha^2}{45m^4}B^2\sin^2(\theta)$ where θ is the angle between the wave vector and the magnetic field, while the latter has an index of refraction of $1 + \frac{14\alpha^2}{45m^4}B^2\sin^2(\theta)$.

Interference Colour Chart

An interference colour chart, first developed by Auguste Michel-Lévy, is an optical mineralogy tool to identify minerals in thin section using a petrographic microscope. With a known thickness of the thin section, minerals have specific and predictable colours in cross-polarized light, and this chart can help identify minerals. The colours are produced by the difference in speed in the fast and slow rays, also known as birefringence.

When using the chart, it is important to remember these tips:

- Isotropic and opaque (metallic) minerals do not apply.
- Rotate the stage in order to get the maximum colour, and therefore, the maximum birefringence.
- Each mineral, depending on the orientation, may not exhibit the maximum birefringence. It is important to try and find multiple like minerals in order to get the best value of birefringence.
- Uniaxial minerals can look isotropic (always extinct) if the mineral is cut perpendicular to the optic axis.

Pleochroism

Pleochroism is an optical phenomenon in which mineral grains within a rock appear to be different colours when observed at different angles under a polarizing petrographic microscope.

Background

Pleochroism is caused by the double refraction of light by a mineral. Light of different polarizations is bent different amounts by the crystal, and therefore follows different paths through the crystal. The components of a divided light beam follow different paths within the mineral and travel at different speeds, and each path will absorb different colours of light. When the mineral is observed at some angle, light following some combination of paths and polarizations will be present, each of which will have had light of different colours absorbed. At another angle, the light passing through the crystal will be composed of another combination of light paths and polarizations, each with their own colour. The light passing through the mineral will therefore have different colours when it is viewed from different angles, making the stone seem to be of different colours.

Tetragonal, trigonal and hexagonal minerals can only show two colours and are called dichroic. Orthorhombic, monoclinic and triclinic crystals show three and are trichroic. Isometric minerals cannot exhibit pleochroism. Tourmaline is notable for exhibiting strong pleochroism. Gems are sometimes cut and set either to display pleochroism or to hide it, depending on the colours and their attractiveness.

In Mineralogy

Pleochroism is an extremely useful tool in mineralogy for mineral identification, since minerals that are otherwise very similar often have very different pleochroic colour schemes. In such cases, a thin section of the mineral is used and examined under polarized transmitted light with a petrographic microscope.

Extinction Angle

Extinction is a term used in optical mineralogy and petrology, which describes when cross-polarized light dims, as viewed through a thin section of a mineral in a petrographic microscope. An isotropic minerals, opaque (metallic) minerals, or amophous materials (glass) show no light (i.e. constant extinction). Anisotropic minerals will show one extinction for each 90 degrees of stage rotation.

The Extinction angle is the measure between the cleavage direction or habit of a mineral and the extinction. To find this, simply line up the cleavage lines/long direction with one of the cross hairs in the microscope, and turn the mineral until the extinction occurs. The number of degrees the stage was rotated is the extinction angle, between 0-89 degrees. 90 degrees would be considered zero degrees, and is known as parallel extinction. Inclined extinction is a measured angle between 1-89 degrees. Minerals with two cleavages can have two extinction angles, and minerals in which the multiple angles are the same are called symmetrical extinction. Minerals that have no cleavage or elongation can not have an extinction angle.

Minerals with undulose extinction, solid solution/zonation, or other factors (e.g. Bird's eye extinction in mica) that may inhibit this measure and may be more difficult to use.

Conoscopic Interference Pattern

A conoscopic interference pattern or interference figure is the best way to determine if a mineral is uniaxial or biaxial and also for determining optic sign in optical mineralogy. The observed interference figure essentially shows all possible birefringence colours at once, including the extinctions (in dark bands called isogyres).

In optical mineralogy a petrographic microscope and cross-polarized light are often used to view the interference pattern. This is done by placing a Bertrand lens (Emile Bertrand, 1878) between a high-power microscope objective and the eyepiece. The microscope's condenser is brought up close underneath the specimen to produce a wide divergence of polarized rays through a small point. There are many other techniques used to observe the interference pattern.

A uniaxial mineral will show a typical 'Maltese' cross shape and its isogyres, which will revolve/orbit around a projection of the optical axis as the stage is rotated.

A biaxial mineral will typically show a saddle-shaped figure (with one isogyre thicker than the other, typically) that will often morph into to curved isogyres with rotation of the stage. The difference in these curved isogyres is known as the "2V" angle. In minerals that have far-off-centre optic axes, only one part of the above sequence may be seen.

A Michel-Levy Chart is often used in conjunction with the interference pattern to determine useful information that aids in the identification of minerals.

Simplified Discussion

A Conoscopic interference pattern or interference figure is the best way to determine if a mineral has one direction of single refraction or two directions of single refraction. (Herein after referred to as Birefringence) It is also used to tell which ray of light is faster in birefraction. The observed interference figure essentially shows all possible birefringence colours at once, including the extinctions. (in dark bands called isogyres)

Light as we see it is seemingly white, but when it is viewed through a mineral we can see that it is separated into a body of colours and each individual colour represents a particular wavelength. As a wavelength each colour has a different speed. Also minerals can inhibit light, limiting lights movement; So when light passes through a mineral it is separated into many different wavelengths, sometimes getting trapped in certain places resulting in a black colour. For example: When a mineral is observed through a petrographic microscope with a cross-polarized light you will see multiple colours and some darker lines within the mineral. (The trapped light referred to as isogyres) In summation, Conoscopic interference patterns are essentially what we use to see how light reacts to certain minerals, and how we can see how much or how little light is inhibited by said mineral.

Becke line test

The Becke Line Test is a technique in optical mineralogy that helps determine the relative refractive index of two materials. It is done by lowering the stage (increasing the focal distance) of the petrographic microscope and observing which direction the light appears to 'move' toward. This movement will always go into the material of higher refractive index. Typically, this is done by comparing 1) two (or more) minerals, 2) a mineral versus thin section epoxy (such as Canada Balsam which has a moderate refractive index of 1.54), and/or 3) a mineral versus an oil of known refractive index (as in Oil immersion studies).

The method was developed by Friedrich Johann Karl Becke (1855-1931).

Optical Relief

Optical relief (usually noted as simply relief) is a concept in optical mineralogy which refers to the degree in which mineral grains stand out from the mounting medium, usually either oil with a known refractive index or Canada Balsam. Relief is an important part of the Becke line test.

Magnitude

Minerals that stand out significantly (have a difference in refractive index of.12 or more) have high or strong relief, and will have very sharp boundaries between itself and the material it is next to. Intermediate is .04 to .12, and low or weak is less than .04. Low relief materials have boundaries that are hard to distinguish from each other.

Polarity

Positive relief refers to a mineral that stands out higher than the medium, and negative relief is a mineral that appears to "sink in".

Wave Plate

A wave plate or retarder is an optical device that alters the polarization state of a light wave travelling through it.

Operation

A wave plate works by shifting the phase between two perpendicular polarization components of the light wave. A typical wave plate is simply a birefringent crystal with a carefully chosen orientation and thickness. The crystal is cut so that the extraordinary axis or "optic axis" is parallel to the surfaces of the plate. Light polarized along this axis travels through the crystal at a different speed than light with the perpendicular polarization, creating a phase difference. When the extraordinary index is smaller than the ordinary index, as in calcite, the extraordinary axis is called the "fast axis" and the perpendicular direction in the plane of the surfaces is called the "slow axis".

Depending on the thickness of the crystal, light with polarization components along both axes will emerge in a different polarization state. The wave plate is characterized by the amount of relative phase, Γ, that it imparts on the two components, which is related to the birefringence Δn and the thickness L of the crystal by the formula

$$\Gamma = \frac{2\pi \Delta n L}{\lambda_0}$$

where λ_0 is the vacuum wavelength of the light. For instance a quarter-wave plate creates a quarter-wavelength phase shift and can change linearly polarized light to circular and vice versa. This is done by adjusting the plane of the incident light so that it makes 45° angle with the fast axis. This gives ordinary and extraordinary waves with equal amplitude.

The other common type of wave plate is a half-wave plate, which retards one polarization by half a wavelength, or 180 degrees. This type of wave plate changes the polarization direction of linear polarized light. Wave plates in general as well as polarizers can be described using the Jones matrix formalism, which uses a vector to represent the polarization state of light and a matrix to represent the linear transformation of a wave plate or polarizer.

Although the birefringence Δn may vary slightly due to dispersion, this is negligible compared to the variation in phase difference according to the wavelength of the light due to the fixed path difference (λ_0 in the denominator in the above equation). Waveplates are thus manufactured to work for a particular range of wavelengths. The phase variation can be minimized by stacking two waveplates that differ by a tiny amount in thickness back-to-back, with the slow axis of one along the fast axis of the other. With this configuration, the relative phase imparted can be, for the case of a quarter-wave plate, one-fourth a wavelength rather than three-fourths or one-fourth plus an integer. This is called a zero-order wave plate.

For a single wave plate changing the wavelength of the light introduces a linear error in the phase. Tilt of the wave plate enters via 1/cos into the path length and thus only quadratically into the phase. For the extraordinary polarization the tilt also changes the refractive index to the ordinary via cos, so combined with the path length, the phase shift for the extraordinary light due to tilt is zero.

A polarization independent phase shift of zero order needs a plate with thickness of one wavelength. For Calcite the refractive index changes in the first decimal place, so that a true zero order plate is ten times as thick as one wavelength. For quartz and magnesium fluoride the refractive index changes in the second decimal place and true zero order plates are common for wave-lengths above 1 μm.

History

William Nicol, whose name is associated with the creation of the Nicol prism, seems to have been the first to prepare thin slices of mineral substances, and his methods were applied by Henry Thronton Maire Witham (1831) to the study of plant petrifactions. This method, of such far-reaching importance in petrology, was not at once made use of for the systematic investigation of rocks, and it was not until

1858 that Henry Clifton Sorby pointed out its value. Meanwhile the optical study of sections of crystals had been advanced by Sir David Brewster and other physicists and mineralogists and it only remained to apply their methods to the minerals visible in rock sections.

Sections

A rock-section should be about one-thousandth of an inch (30 micrometres) in thickness, and is relatively easy to make. A thin splinter of the rock, about 1 centimetre may be taken; it should be as fresh as possible and free from obvious cracks. By grinding it on a plate of planed steel or cast iron with a little fine carborundum it is soon rendered flat on one side and is then transferred to a sheet of plate glass and smoothed with the very finest emery till all minute pits and roughnesses are removed and the surface is a uniform plane. The rock-chip is then washed, and placed on a copper or iron plate which is heated by a spirit or gas lamp. A microscopic glass slip is also warmed on this plate with a drop of viscous natural Canada balsam on its surface. The more volatile ingredients of the balsam are dispelled by the heat, and when that is accomplished the smooth, dry, warm rock is pressed firmly into contact with the glass plate so that the film of balsam intervening may be as thin as possible and free from air-bubbles. The preparation is allowed to cool and then the rock chip is again ground down as before, first with carborundum and, when it becomes transparent, with fine emery till the desired thickness is obtained. It is then cleaned, again heated with a little more balsam, and covered with a cover glass. The labour of grinding the first surface may be avoided by cutting off a smooth slice with an iron disk armed with crushed diamond powder. A second application of the slitter after the first face is smoothed and cemented to the glass will in expert hands leave a rock-section so thin as to be already transparent. In this way the preparation of a section may require only twenty minutes.

Microscope

The microscope employed is usually one which is provided with a rotating stage beneath which there is a polarizer, while above the objective or the eyepiece an analyser is mounted; alternatively the stage may be fixed and the polarizing and analysing prisms may be capable of simultaneous rotation by means of toothed wheels and a connecting-rod. If ordinary light and not polarized light is desired, both prisms may be withdrawn from the axis of the instrument; if the polarizer only is inserted the light transmitted is plane polarized; with both prisms in position the slide is viewed in cross-polarized light, also known as "crossed nicols." A microscopic rock-section in ordinary light, if a suitable magnification (say 30) be employed, is seen to consist of grains or crystals varying in colour, size and shape.

Characters of Minerals

Some minerals are colourless and transparent (quartz, calcite, feldspar, muscovite, etc.), others are yellow or brown (rutile, tourmaline, biotite), green (diopside, hornblende, chlorite), blue (glaucophane), pink (garnet), etc. The same mineral may present a variety of colours, in the same or different rocks, and these colours may be arranged in zones parallel to the surfaces of the crystals. Thus tourmaline may be brown, yellow, pink, blue, green, violet, grey, or colourless, but every mineral has one or more characteristic, most common tints. The shapes of the crystals determine in a general way the outlines of the sections of them presented on the slides. If the mineral has one or more good cleavages they will be indicated by systems of cracks. The refractive index is also clearly shown by the appearance of the section, which are rough, with well-defined borders if they have a much stronger refraction than the medium in which they are mounted. Some minerals decompose readily and become turbid and semi-transparent (e.g. feldspar); others remain always perfectly fresh and clear (e.g. quartz), others yield characteristic secondary products (such as green chlorite after biotite). The inclusions in the crystals (both solid and fluid) are of great interest; one mineral may enclose another, or may contain spaces occupied by glass, by fluids or by gases.

Microstructure

Lastly the structure of the rock, that is to say, the relation of its components to one another, is usually clearly indicated, whether it be fragmented or massive; the presence of glassy matter in contradistinction to a completely crystalline or "holo-crystalline" condition; the nature and origin of organic fragments; banding, foliation or lamination; the pumiceous or porous structure of many lavas; these and many other characters, though often not visible in the hand specimens of a rock, are rendered obvious by the examination of a microscopic section. Many refined methods of observation may be introduced, such as the measurement of the size of the elements of the rock by the help of micrometers; their relative proportions by means of a glass plate ruled in small squares; the angles between cleavages or faces seen in section by the use of the rotating graduated stage, and the estimation of the refractive index of the mineral by comparison with those of different mounting media.

Pleochroism

Further information is obtained by inserting the polarizer and rotating the section. The light vibrates now only in one plane, and in passing through doubly refracting crystals in the slide, is, speaking generally, broken up into rays, which vibrate at right angles to one another. In many coloured minerals such as biotite, hornblende, tourmaline, chlorite, these two rays have different colours, and when a section containing any of these minerals is rotated the change of colour is often very striking. This property, known as "pleochroism" is of great value in the determination of rock-making minerals.

Pleochroism is often especially intense in small spots which surround minute enclosures of other minerals, such as zircon and epidote, these are known as "pleochroic halos."

Double Refraction

If the analyser be now inserted in such a position that it is crossed relatively to the polarizer the field of view will be dark where there are no minerals, or where the light passes through isotropic substances such as glass, liquids and cubic crystals. All other crystalline bodies, being doubly refracting, will appear bright in some position as the stage is rotated. The only exception to this rule is provided by sections which are perpendicular to the optic axes of birefringent crystals; these remain dark or nearly dark during a whole rotation, and as will be seen later, their investigation is of special importance.

Extinction

The doubly refracting mineral sections, however, will in all cases appear black in certain positions as the stage is rotated. They are said to go "extinct" when this takes place. If we note these positions we may measure the angle between them and any cleavages, faces or other structures of the crystal by means of the rotating stage. These angles are characteristic of the system to which the mineral belongs and often of the mineral species itself. To facilitate measurement of extinction angles various kinds of eyepieces have been devised, some having a stereoscopic calcite plate, others with two or four plates of quartz cemented together; these are often found to give more exact results than are obtained by observing merely the position in which the mineral section is most completely dark between crossed nicols.

The mineral sections when not extinguished are not only bright but are coloured and the colours they show depend on several factors, the most important of which is the strength of the double refraction. If all the sections are of the same thickness as is nearly true of well-made slides, the minerals with strongest double refraction yield the highest polarization colours.

The order in which the colours are arranged in what is known as Newton's scale, the lowest being dark grey, then grey, white, yellow, orange, red, purple, blue and so on. The difference between the refractive indexes of the ordinary and the extraordinary ray in quartz is .009, and in a rock-section about 1/500 of an inch thick this mineral gives grey and white polarization colours; nepheline with weaker double refraction gives dark grey; augite on the other hand will give red and blue, while calcite with the stronger double refraction will appear pinkish or greenish white. All sections of the same mineral, however, will not have the same colour; it was stated above that sections perpendicular to an optic axis will be nearly black, and, in general, the more nearly any section approaches this direction the lower its polarization colours will be. By taking the average, or the highest colour given by any mineral, the relative value of its double refraction can be estimated; or if the thickness of the section be precisely known the difference between the two refractive indexes can be ascertained. If the slides be thick the colours will be on the whole higher than in thin slides.

It is often important to find out whether of the two axes of elasticity (or vibration traces) in the section is that of greater elasticity (or lesser refractive index). The quartz wedge or selenite plate enables us to do this. Suppose a doubly refracting mineral section so placed that it is "extinguished"; if now is rotated through 45 degrees it will be brightly illuminated. If the quartz wedge be passed across it so that the long axis of the wedge is parallel to the axis of elasticity in the section the polarization colours will rise or fall. If they rise the axes of greater elasticity in the two minerals are parallel; if they sink the axis of greater elasticity in the one is parallel to that of lesser elasticity in the other. In the latter case by pushing the wedge sufficiently far complete darkness or compensation will result. Selenite wedges, selenite plates, mica wedges and mica plates are also used for this purpose. A quartz wedge also may be calibrated by determining the amount of double refraction in all parts of its length. If now it be used to produce compensation or complete extinction in any doubly refracting mineral section, we can ascertain what is the strength of the double refraction of the section because it is obviously equal and opposite to that of a known part of the quartz wedge.

A further refinement of microscopic methods consists of the use of strongly convergent polarized light (konoscopic methods). This is obtained by a wide angled achromatic condenser above the polarizer, and a high power microscopic objective. Those sections are most useful which are perpendicular to an optic axis, and consequently remain dark on rotation. If they belong to uniaxial crystals they show a dark cross or convergent light between crossed nicols, the bars of which remain parallel to the wires in the field of the eyepiece. Sections perpendicular to an optic axis of a biaxial mineral under the same conditions show a dark bar which on rotation becomes curved to a hyperbolic shape. If the section is perpendicular to a "bisectrix" a black cross is seen which on rotation opens out to form two hyperbolas, the apices of which are turned towards one another. The optic axes emerge at the apices of the hyperbolas and may be surrounded by coloured rings, though owing to the thinness of minerals in rock sections these are only seen when the double refraction of the mineral is strong. The distance between the axes as seen in the field of the microscope depends partly on the axial angle of the crystal and partly on the numerical aperture of the objective. If it is measured by means of eye-piece micrometer, the optic axial angle of the mineral can be found by a simple calculation. The quartz wedge, quarter mica plate or selenite plate permit the determination of the positive or negative character of the crystal by the changes in the colour or shape of the figures observed in the field. These operations are precisely similar to those employed by the mineralogist in the examination of plates cut from crystals. It is sufficient to point out that the petrological microscope in its modern development is an optical instrument of great precision, enabling us to determine physical constants of crystallized substances as well as serving to produce magnified images like the ordinary microscope. A great variety of accessory apparatus has been devised to fit it for these special uses.

Examination of Rock Powders

Although rocks are now studied principally in microscopic sections the investigation of fine crushed rock powders, which was the first branch of microscopic petrology to receive attention, is by no means discontinued. The modern optical methods are perfectly applicable to transparent mineral fragments of any kind. Minerals are almost as easily determined in powder as in section, but it is otherwise with rocks, as the structure or relation of the components to one another, which is an element of great importance in the study of the history and classification or rocks, is almost completely destroyed by grinding them to powder.

Formation Environments

The environments of mineral formation and growth are highly varied, ranging from slow crystallization at the high temperature and pressures of igneous melts deep within the Earth's crust to the low temperature precipitation from a saline brine at the Earth's surface.

Various possible methods of formation include:

- sublimation from volcanic gases
- deposition from aqueous solutions and hydrothermal brines
- crystallization from an igneous magma or lava
- recrystallization due to metamorphic processes and metasomatism
- crystallization during diagenesis of sediments
- formation by oxidation and weathering of rocks exposed to the atmosphere or within the soil environment.

Descriptive Mineralogy

Descriptive mineralogy summarizes results of studies performed on mineral substances. It is the scholarly and scientific method of recording the identification, classification, and categorization of minerals, their properties, and their uses. Classifications for descriptive mineralogy includes:

Mineral Habits

In mineralogy a mineral habit is the characteristic external form or appearance of a mineral.

Several examples of mineral habits are: prismatic, dentric, bladed, acicular, massive, and reniform.

Acicular (Crystal Habit)

Acicular, in mineralogy, refers to a crystal habit composed of a radiating mass of slender, needle-like crystals. Minerals with this habit tend to be fragile and complete, undamaged specimens can be uncommon.

Examples

Some minerals with an acicular habit include mesolite, natrolite, malachite, gypsum, rutile, and brochantite.

Differences From Other Habits

Some minerals like creedite form prismatic crystals that appear to be acicular, but are instead prismatic in a bladelike form; these can be told apart by the fact that all prismatic crystals are less sharp, sometimes are tipped with a pyramidal shape, and keep a standard cross-section shape with straight edges.

Acicular crystals differ from fibrous crystals in their thickness; crystals with a fibrous habit are much thinner, sometimes to the point of being flexible like hair, while acicular crystals are thicker and rigid.

Botryoidal

A botryoidal texture or mineral habit is one in which the mineral has a globular external form resembling a *bunch of grapes* as derived from the Greek.

This is a common form for many minerals particularly hematite where it is the classically recognized shape. It is also a common form of goethite, smithsonite, fluorite and malachite. This includes chrysocolla.

Each sphere (grape) in a botryoidal mineral is smaller than that of a reniform mineral, and much smaller than that of a mamillary mineral. Botryoidal minerals form when many nearby nuclei, specks of sand, dust, or other particles, are present. Layers of mineral material are deposited radially around the nuclei. As more material is deposited, the spheres grow larger and eventually overlap with those that are nearby. These nearby spheres are then fused together to form the botryoidal cluster.

Druse (Geology)

In geological usage druse or druzy is a coating of fine crystals on a rock fracture surface, vein or within a vug or geode.

Occurrence

Druse occurs worldwide, the most common is perhaps quartz druse within voids in chert or agates. Garnet, calcite, dolomite and a variety of minerals may occur as druse coatings.

Euhedral

Euhedral crystals are those that are well-formed with sharp, easily recognised faces. Normally, crystals do not form smooth faces or sharp crystal outlines. Many crystals grow from cooling liquid magma. As magma cools, the crystals grow and eventually touch each other, preventing crystal faces from forming properly or at all.

However, when snowflakes crystallize, they do not touch each other. Thus, snowflakes form euhedral, six-sided twinned crystals. In rocks, the presence of euhedral crystals may signify that they formed early in the crystallization of a magma or perhaps crystallized in a cavity or vug, without hindrance from other crystals.

Etymology: Euhedral is derived from the Greek *eu* meaning *true* and *hedron* meaning *shape*.

Prism (Geology)

In geology and petrology, a prism is a long, narrow, wedge-shaped sedimentary body with a width to thickness ratio greater than 5 to 1 but less than 50 to 1, and a length 1.5 to 3 times its width. These types of sediments are typically formed during orogenic deformation. For example, the arkoses found in fault troughs.

"Prismatic" is also type of mineral habit (appearance of a crystal). Prismatic minerals have crystals that show a uniform cross-section.

Petrographic Microscope

A petrographic microscope is a type of optical microscope used in petrology and optical mineralogy to identify rocks and minerals in thin sections. The microscope is used in optical mineralogy and petrography, a branch of petrology which focuses on detailed descriptions of rocks. The method is called "polarized light microscopy" (PLM). Depending on the grade of observation required, petrological microscopes are derived from conventional brightfield microscopes of similar basic capabilities by:

- adding a polarizer filter to the light path beneath the sample slide
- replacing the normal stage with a circular rotating stage (typically graduated with vernier scales for reading orientations to better than 1 degree of arc)
- adding a second rotatable and removable polarizer filter, called the analyser, to the light path between objective and eyepiece

- adding a Phase telescope, also known as a Betrand Lens, which allows the viewer to see conoscopic interference patterns
- adding a slot for insertion of wave plates.

Petrographic microscopes are constructed with optical parts that do not add unwanted polarizing effects due to strained glass, or polarization by reflection in prisms and mirrors.

These special parts add to the cost and complexity of the microscope. However, a "simple polarizing" microscope is easily made by adding inexpensive polarizing filters to a standard biological microscope, often with one in a filter holder beneath the condenser, and a second inserted beneath the head or eyepiece. These might be sufficient for many non-quantitative purposes.

The two filters of the petrographic microscope have their polarizing planes oriented perpendicular to one another. When only an isotropic material such as air, water, or glass exists between the filters, all light is blocked. However, most crystalline materials and minerals change the polarizing light directions, which allows some of the altered light to pass through the analyser to the viewer. Using one polarizer allows for looking at the slide in plane polarized light, while using two allows for analysis under cross polarized light.

A particular light pattern on the upper lens surface of the objectives is created as an conoscopic interference pattern characteristic of uniaxial and biaxial minerals, and produced with convergent polarized light. To observe the interference figure, true petrographic microscopes usually include an accessory called a Bertrand lens, which focuses and enlarges the figure. It is also possible to remove an eyepiece lens to make a direct observation of the objective lens surface.

In addition to modifications of the microscope's optical system, petrographic microscopes allow for the insertion of specially-cut oriented filters of biaxial minerals (named the Quartz Wedge, quarter-wave mica plate and half-wave mica plate), into the optical train between the polarizers to identify positive and negative birefringence, and in extreme cases, the mineral order when needed.

Optic Axis of a Crystal

The optic axis of a crystal is the direction in which a ray of transmitted light suffers no birefringence (double refraction). Due to the internal structure of the crystal (the specific structure of the crystal lattice, the form of atoms or molecules of its components), light propagates along the optical axis differently than in other directions. Light propagating along the optical axis of a uniaxial crystal (e.g. calcite, quartz), has no unusual results. Light propagates along that axis with a speed independent of its polarization. If the light beam is not parallel to the optical axis, then the beam is split into two rays (the ordinary and extraordinary) when passing through the crystal. These rays will be mutually perpendicularly polarized.

The optic axis of a crystal is a direction rather than a single line. If a ray in this direction suffers no birefringence, neither will all parallel rays. A crystal with only one optic axis is called an uniaxial crystal. Crystals are classed according to the number of optic axes (uniaxial, biaxial). An uniaxial crystal is isotropic within the plane orthogonal to the optical axis of the crystal.

The refractive index of the ordinary ray is constant for any direction in the crystal. The refractive index of the extraordinary ray is variable and depends on the direction. Non-crystalline materials have no double refraction and thus, no optic axis. Some solid materials under specific conditions can demonstrate double refraction and optic axes.

Thin Section

In optical mineralogy and petrography, a thin section is a laboratory preparation of a rock, mineral, soil, pottery, bones, even metal sample for use with a polarizing petrographic microscope, electron microscope and electron microprobe.

A thin sliver of rock is cut from the sample with a diamond saw and ground optically flat. It is then mounted on a glass slide and then ground smooth using progressively finer abrasive grit until the sample is only 30 μm thick. The method involved using the Michel-Lévy interference colour chart. Typically quartz is used as the gauge to determine thickness as it is one of the most abundant minerals.

When placed between two polarizing filters set at right angles to each other, the optical properties of the minerals in the thin section alter the colour and intensity of the light as seen by the viewer. As different minerals have different optical properties, most rock forming minerals can be easily identified. Plagioclase for example can be seen in the photo on the right as a clear mineral with multiple parallel twinning planes. The large blue-green minerals are clinopyroxene with some exsolution of orthopyroxene.

Thin sections are prepared in order to investigate the optical properties of the minerals in the rock. This work is a part of petrology and helps to reveal the origin and evolution of the parent rock.

Ultra-thin Sections

Fine-grained rocks, particularly those containing minerals of high birefringence, such as calcite, are sometimes prepared as ultra-thin sections.

An ordinary 30 μm thin section is prepared but the slice of rock is attached to the glass slide using a soluble cement such as canada balsam (soluble in ethanol) to allow both sides to be worked on. The section is then polished on both sides using a fine diamond paste until it has a thickness in the range of 2-12 μm.

This technique has been used to study the microstructure of fine-grained carbonates such as the Lochseitenkalk mylonite in which the matrix grains are less than 5 μm in size.

This method is also sometimes used in the preparation of mineral and rock specimens for transmission electron microscopy and allows greater accuracy in comparing features using both optical and electron imaging.

Undulose Extinction

Undulose extinction or undulatory extinction is a geological term referring to the type of extinction that occurs in certain minerals when examined in thin section under cross polarized light.

As the microscope stage is rotated, individual mineral grains appear black when the polarization due to the mineral prevents any light from passing through. If a mineral is deformed plastically by dislocation processes without recovery, strain builds up within the crystal lattice causing it to warp. This means that different parts reach extinction at slightly different angles, giving the crystal an irregular, mottled look.

Undulose extinction is very common in quartz, so much so that it is often used as a diagnostic feature of that mineral, and feldspar of various sorts, but is possible in almost any mineral.

It is possible to identify the presence of a porphyroblast before the main deformation event in a metamorphic reaction, if a porphyroblast displays clear undulose extinction it is a clear sign that it existed before the main rock was created.

Bird s Eye Maple (Mineral Property)

Bird's eye maple, or bird's eye extinction, is a specific type of extinction exhibited by minerals of the mica group under cross polarized light (sometimes called the optical analyser). It gives the mineral a pebbly appearance as it passes into extinction. This is caused when the grinding tools used to create petrographic slides of precise widths alter the alignment of the previously perfect basal cleavage planes which split micas up into its characteristic thin sheets. The resulting, slightly roughened surface alters the extinction angle of various parts of the crystal lattice, leading to this type of extinction. Since it is not a natural feature of the mineral, bird's eye maple is not observed in all mica crystals, nor from all angles, but it is quite common, and is used as a diagnostic feature for micas.

Common micas which exhibit this include biotite (and the magnesium end-member phlogopite) and muscovite.

Chapter 4

Mineral Physics

Mineral physics is the science of materials that compose the interior of planets, particularly the Earth. It overlaps with petrophysics, which focuses on whole-rock properties. It provides information that allows interpretation of surface measurements of seismic waves, gravity anomalies, geomagnetic fields and electromagnetic fields in terms of properties in the deep interior of the Earth. This information can be used to provide insights into plate tectonics, mantle convection, the geodynamo and related phenomena.

Laboratory work in mineral physics require high pressure measurements. The most common tool is a diamond anvil cell, which uses diamonds to put a small sample under pressure that can approach the conditions in the Earth's interior.

Creating High Pressures

The pace of progress in mineral physics has been determined, to a large extent, by the technology for reproducing the high pressures and temperatures in the Earth's interior. The most common tools for achieving this have been:

Shock Compression

Many of the pioneering studies in mineral physics involved explosions or projectiles that subjected a sample to a shock. For a brief time interval, the sample is under pressure as the shock wave passes through. Pressures as high as any in the Earth have been achieved by this method. However, the method has some disadvantages. The pressure is very non-uniform and is not adiabatic, so the pressure wave heats the sample up in passing. The conditions of the experiment must be interpreted in terms of a set of pressure-density curves called the Hugoniot curves.

Multi-anvil Press

Multi-anvil presses involve an arrangement of anvils to concentrate pressure from a press onto a sample. Unlike shock compression, the pressure exerted is steady, and the sample can be heated using a furnace. Pressures equivalent to depths of 700 km and temperatures of 1500°C can be attained. The apparatus is very bulky and cannot achieve pressures like those in the diamond anvil cell (below), but they can handle much larger samples that can be examined after the experiment.

Diamond Anvil Cell

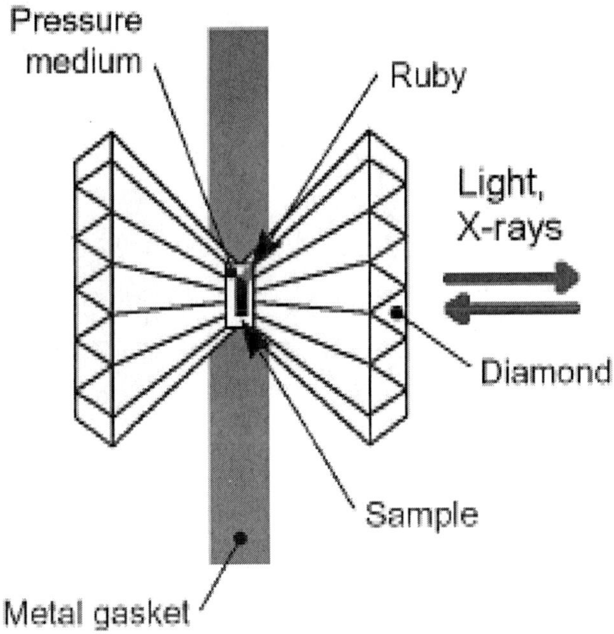

Figure 1 : Schematics of the Core of a Diamond Anvil cell. The Diamond size is a few Millimeters at Most

The diamond anvil cell is a small table-top device for concentrating pressure. It can compress a small (sub-millimetre sized) piece of material to extreme pressures, which can exceed 3,000,000 atmospheres (300 gigapascals). This is beyond the pressures at the centre of the Earth. The concentration of pressure at the tip of the diamonds is possible because of their hardness, while their transparency and high electrical conductivity allow a variety of probes can be used to examine the state of the sample. The sample can be heated to thousands of degrees.

Properties of Materials

Equations of State

To deduce the properties of minerals in the deep Earth, it is necessary to know how their density varies with pressure and temperature. Such a relation is called an equation of state (EOS). A simple example of an EOS that is predicted by the Debye model for harmonic lattice vibrations is the Mie-Grünheisen equation of state:

$$\left(\frac{dP}{dT}\right)=\frac{\gamma_D}{V}C_V,$$

where C_V is the heat capacity and γ_D is the Debye gamma. The latter is one of many Grünheisen parameters that play an important role in high-pressure physics. A more realistic EOS is the Birch–Murnaghan equation of state.

Interpreting Seismic Velocities

Inversion of seismic data give profiles of seismic velocity as a function of depth. These must still be interpreted in terms of the properties of the minerals. A very useful heuristic was discovered by Francis Birch: plotting data for a large number of rocks, he found a linear relation of the compressional wave velocity v_p of rocks and minerals of a constant average atomic weight $\bar{M}$ with density ρ:

$$v_p = a\bar{M} + b\rho.$$

This makes it possible to extrapolate known velocities for minerals at the surface to predict velocities deeper in the Earth.

Other Physical Properties

- Viscosity
- Creep (deformation)
- Melting
- Electrical conduction and other transport properties

Viscosity

Viscosity is a measure of the resistance of a fluid which is being deformed by either shear or tensile stress. In everyday terms (and for fluids only), viscosity is "thickness" or "internal friction". Thus, water is "thin", having a lower viscosity, while honey is "thick", having a higher viscosity. Put simply, the less viscous the fluid is, the greater its ease of movement (fluidity). Viscosity describes a fluid's internal resistance to flow and may be thought of as a measure of fluid friction. For example, high-viscosity felsic magma will create a tall, steep stratovolcano, because it cannot flow far before it cools, while low-viscosity mafic lava will create a wide, shallow-sloped shield volcano. All real fluids (except superfluids) have some resistance to stress and therefore are viscous, but a fluid which has no resistance to shear stress is known as an ideal fluid or inviscid fluid.

The study of flowing matter is known as rheology, which includes viscosity and related concepts.

Etymology

The word "viscosity" is derived from the Latin "viscum alba", meaning white mistletoe. A viscous glue called birdlime was made from mistletoe berries and was used for lime-twigs to catch birds.

Properties and Behaviour

Overview

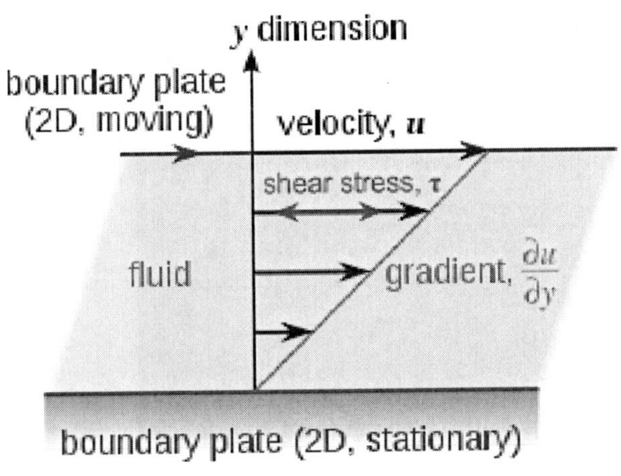

Figure 2: Laminar shear of fluid between two plates. Friction between the fluid and the moving boundaries causes the fluid to shear. The force required for this action is a measure of the fluid's viscosity. This type of flow is known as a Couette flow.

In general, in any flow, layers move at different velocities and the fluid's viscosity arises from the shear stress between the layers that ultimately opposes any applied force. The relationship between the shear stress and the velocity gradient can be obtained by considering two plates closely spaced at a distance y, and separated by a homogeneous substance. Assuming that the plates are very large, with a large area A, such that edge effects may be ignored, and that the lower plate is fixed, let a force F be applied to the upper plate. If this force causes the substance between the plates to undergo shear flow with a velocity gradient u/y (as opposed to just shearing elastically until the shear stress in the substance balances the applied force), the substance is called a fluid.

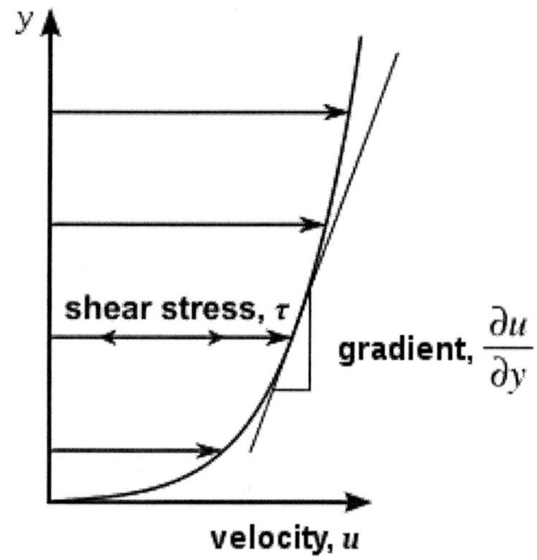

Figure 3 : Laminar shear, the non-constant gradient, is a result of the geometry the fluid is flowing through (e.g. a pipe).

The applied force is proportional to the area and velocity gradient in the fluid and inversely proportional to the distance between the plates. Combining these three relations results in the equation:

$$F = \mu A \frac{u}{y},$$

where μ is the proportionality factor called *dynamic viscosity*.

This equation can be expressed in terms of shear stress $\tau = \frac{F}{A}$. Thus as expressed in differential form by Isaac Newton for straight, parallel and uniform flow, the shear stress between layers is proportional to the velocity gradient in the direction perpendicular to the layers:

$$\tau = \mu \frac{\partial u}{\partial y}$$

Hence, through this method, the relation between the shear stress and the velocity gradient can be obtained.

Note that the *rate of shear deformation* is $\frac{u}{y}$ which can be also written as a *shear velocity*, $\frac{du}{dy}$.

James Clerk Maxwell called viscosity *fugitive elasticity* because of the analogy that elastic deformation opposes shear stress in solids, while in viscous fluids, shear stress is opposed by *rate* of deformation.

Types of Viscosity

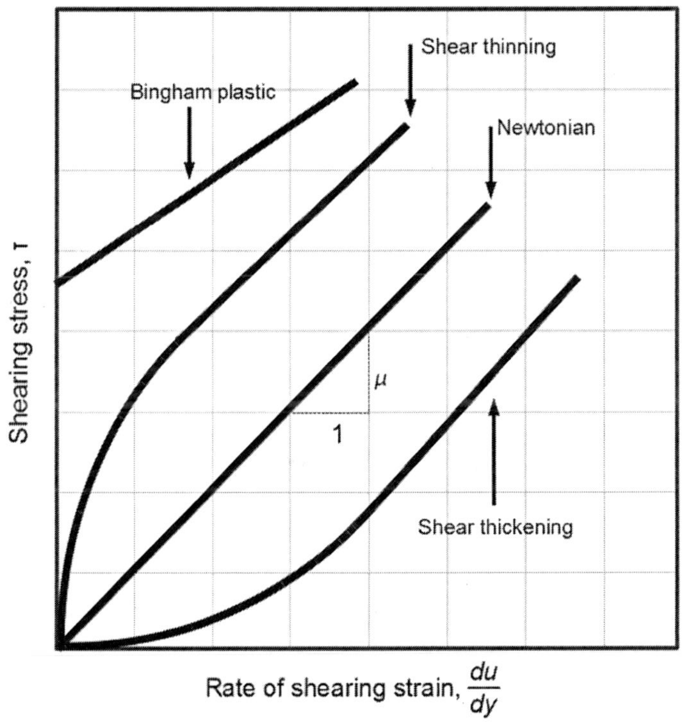

Figure 4: Viscosity, the Slope of Each line, Varies Among Materials

Newton's law of viscosity, given above, is a constitutive equation (like Hooke's law, Fick's law, Ohm's law). It is not a fundamental law of nature but an approximation that holds in some materials and fails in others. Non-Newtonian fluids exhibit a more complicated relationship between shear stress and velocity gradient than simple linearity. Thus there exist a number of forms of viscosity:

- Newtonian: fluids, such as water and most gases which have a constant viscosity.
- Shear thickening: viscosity *increases* with the rate of shear.
- Shear thinning: viscosity *decreases* with the rate of shear. Shear thinning liquids are very commonly, but misleadingly, described as thixotropic.
- Thixotropic: materials which become *less* viscous over time when shaken, agitated, or otherwise stressed.
- Rheopectic: materials which become *more* viscous over time when shaken, agitated, or otherwise stressed.
- A Bingham plastic is a material that behaves as a solid at low stresses but flows as a viscous fluid at high stresses.
- A magnetorheological fluid is a type of "smart fluid" which, when subjected to a magnetic field, greatly increases its apparent viscosity, to the point of becoming a viscoelastic solid.

Viscosity Coefficients

Viscosity coefficients can be defined in two ways:

- Dynamic viscosity, also absolute viscosity, the more usual one (typical units Pa·s, Poise, P);
- Kinematic viscosity is the *dynamic viscosity* divided by the density (typical units cm^2/s, Stokes, St).

Viscosity is a tensorial quantity that can be decomposed in different ways into two independent components. The most usual decomposition yields the following viscosity coefficients:

- Shear viscosity, the most important one, often referred to as simply viscosity, describing the reaction to applied shear stress; simply put, it is the ratio between the pressure exerted on the surface of a fluid, in the lateral or horizontal direction, to the change in velocity of the fluid as you move down in the fluid (this is what is referred to as a velocity gradient).
- Volume viscosity (also called bulk viscosity or second viscosity) becomes important only for such effects where fluid compressibility is essential. Examples would include shock waves and sound propagation. It appears in the Stokes' law (sound attenuation) that describes propagation of sound in Newtonian liquid.

- Extensional viscosity, a linear combination of shear and bulk viscosity, describes the reaction to elongation, widely used for characterizing polymers. For example, at room temperature, water has a dynamic shear viscosity of about 1.0×10^{-3} Pa·s and motor oil of about 250×10^{-3} Pa·s.

Viscosity Measurement

Viscosity is measured with various types of viscometers and rheometers. A rheometer is used for those fluids which cannot be defined by a single value of viscosity and therefore require more parameters to be set and measured than is the case for a viscometer. Close temperature control of the fluid is essential to accurate measurements, particularly in materials like lubricants, whose viscosity can double with a change of only 5 °C.

For some fluids, viscosity is a constant over a wide range of shear rates (Newtonian fluids). The fluids without a constant viscosity (non-Newtonian fluids) cannot be described by a single number. Non-Newtonian fluids exhibit a variety of different correlations between shear stress and shear rate.

One of the most common instruments for measuring kinematic viscosity is the glass capillary viscometer.

In paint industries, viscosity is commonly measured with a Zahn cup, in which the efflux time is determined and given to customers. The efflux time can also be converted to kinematic viscosities (centistokes, cSt) through the conversion equations.

Also used in paint, a Stormer viscometer uses load-based rotation in order to determine viscosity. The viscosity is reported in Krebs units (KU), which are unique to Stormer viscometers.

A Ford viscosity cup measures the rate of flow of a liquid. This, under ideal conditions, is proportional to the kinematic viscosity.

Vibrating viscometers can also be used to measure viscosity. These models such as the *Dynatrol* use vibration rather than rotation to measure viscosity.

Extensional viscosity can be measured with various rheometers that apply extensional stress.

Volume viscosity can be measured with an acoustic rheometer.

Apparent viscosity is a calculation derived from tests performed on drilling fluid used in oil or gas well development. These calculations and tests help engineers develop and maintain the properties of the drilling fluid to the specifications required.

Units

Dynamic Viscosity

The usual symbol for dynamic viscosity used by mechanical and chemical engineers — as well as fluid dynamicists — is the Greek letter mu (μ). The symbol ç is also used by chemists, physicists, and the IUPAC.

The SI physical unit of dynamic viscosity is the pascal-second (Pa·s), (equivalent to $N \cdot s/m^2$, or kg/(m·s)). If a fluid with a viscosity of one Pa·s is placed between two plates, and one plate is pushed sideways with a shear stress of one pascal, it moves a distance equal to the thickness of the layer between the plates in one second. Water at 20 °C has a viscosity of 0.001002 Pa·s.

The cgs physical unit for dynamic viscosity is the *poise* (P), named after Jean Louis Marie Poiseuille. It is more commonly expressed, particularly in ASTM standards, as *centipoise* (cP). Water at 20 °C has a viscosity of 1.0020 cP.

1 P = 0.1 Pa·s,

1 cP = 1 mPa·s = 0.001 Pa·s.

Kinematic Viscosity

In many situations, we are concerned with the ratio of the inertial force to the viscous force (i.e. the Reynolds number, $Re = VD/\nu$), the former characterized by the fluid density ρ. This ratio is characterized by the *kinematic viscosity* (Greek letter nu, ν), defined as follows:

$$\nu = \frac{\mu}{\rho}$$

The SI unit of ν is m^2/s. The SI unit of ρ is kg/m^3.

The cgs physical unit for kinematic viscosity is the *stokes* (St), named after George Gabriel Stokes. It is sometimes expressed in terms of *centiStokes* (cSt). In U.S. usage, *stoke* is sometimes used as the singular form.

$1 \text{ St} = 1 \text{ cm}^2 \cdot s^{-1} = 10^{-4} \text{ m}^2 \cdot s^{-1}$.

$1 \text{ cSt} = 1 \text{ mm}^2 \cdot s^{-1} = 10^{-6} \text{m}^2 \cdot s^{-1}$.

Water at 20 °C has a kinematic viscosity of about 1 cSt.

The kinematic viscosity is sometimes referred to as diffusivity of momentum, because it is analogous to diffusivity of heat and diffusivity of mass. It is therefore used in dimensionless numbers which compare the ratio of the diffusivities.

Fluidity

The reciprocal of viscosity is *fluidity*, usually symbolized by $\varphi = 1/\mu$ or $F = 1/\mu$, depending on the convention used, measured in *reciprocal poise* ($cm \cdot s \cdot g^{-1}$), sometimes called the *rhe*. *Fluidity* is seldom used in engineering practice.

The concept of fluidity can be used to determine the viscosity of an ideal solution. For two components *a* and *b*, the fluidity when *a* and *b* are mixed is

$$F \approx \chi_a F_a + \chi_b F_b,$$

which is only slightly simpler than the equivalent equation in terms of viscosity:

$$\mu \approx \frac{1}{\chi_a / \mu_a + \chi_b / \mu_b},$$

where χ_a and χ_b is the mole fraction of component *a* and *b* respectively, and μ_a and μ_b are the components pure viscosities.

Non-standard Units

The Reyn is a British unit of dynamic viscosity.

Viscosity index is a measure for the change of kinematic viscosity with temperature. It is used to characterise lubricating oil in the automotive industry.

At one time the petroleum industry relied on measuring kinematic viscosity by means of the Saybolt viscometer, and expressing kinematic viscosity in units of *Saybolt Universal Seconds* (SUS). Other abbreviations such as SSU (*Saybolt Seconds Universal*) or SUV (*Saybolt Universal Viscosity*) are sometimes used. Kinematic viscosity in centistoke can be converted from SUS according to the arithmetic and the reference table provided in ASTM D 2161.

Molecular Origins

The viscosity of a system is determined by how molecules constituting the system interact. There are no simple but correct expressions for the viscosity of a fluid. The simplest exact expressions are the Green–Kubo relations for the linear shear viscosity or the Transient Time Correlation Function expressions derived by Evans and Morriss in 1985. Although these expressions are each exact in order to calculate the viscosity of a dense fluid, using these relations requires the use of molecular dynamics computer simulations.

Gases

Viscosity in gases arises principally from the molecular diffusion that transports momentum between layers of flow. The kinetic theory of gases allows accurate prediction of the behaviour of gaseous viscosity.

Within the regime where the theory is applicable:

- Viscosity is independent of pressure and
- Viscosity increases as temperature increases.

James Clerk Maxwell published a famous paper in 1866 using the kinetic theory of gases to study gaseous viscosity. To understand why the viscosity is independent of pressure, consider two adjacent boundary layers (A and B) moving with respect to each other. The internal friction (the viscosity) of the gas is determined by the probability a particle of layer A enters layer B with a corresponding transfer of momentum. Maxwell's calculations showed him that the viscosity coefficient is proportional to both the density, the mean free path and the mean velocity of the atoms. On the other hand, the *mean free path* is inversely proportional to the density. So an increase of pressure doesn't result in any change of the viscosity.

Relation to Mean Free Path of Diffusing Particles

In relation to diffusion, the kinematic viscosity provides a better understanding of the behaviour of mass transport of a dilute species. Viscosity is related to shear stress and the rate of shear in a fluid, which illustrates its dependence on the mean free path, λ, of the diffusing particles.

From fluid mechanics, for a Newtonian fluid, the shear stress, τ, on a unit area moving parallel to itself, is found to be proportional to the rate of change of velocity with distance perpendicular to the unit area:

$$\tau = \mu \frac{du_x}{dy}$$

for a unit area parallel to the x-z plane, moving along the x axis. We will derive this formula and show how μ is related to λ.

Interpreting shear stress as the time rate of change of momentum, *p*, per unit area *A* (rate of momentum flux) of an arbitrary control surface gives

$$\tau = \frac{\dot{p}}{A} = \frac{\dot{m}\langle u_x \rangle}{A}.$$

where $\langle u_x \rangle$ is the average velocity along x of fluid molecules hitting the unit area, with respect to the unit area.

Further manipulation will show :

$$\dot{m} = \rho \bar{u} A$$

$$\langle u_x \rangle = \frac{1}{2}\lambda \frac{du_x}{dy},$$

Assuming that molecules hitting the unit area come from all distances between 0 and λ (equally distributed), and that their average velocities change linearly with distance (always true for small enough λ). From this follows:

$$\tau = \underbrace{\frac{1}{2}\rho \bar{u}\lambda}_{\mu} \cdot \frac{du_x}{dy} \Rightarrow \nu = \frac{\mu}{\rho} = \tfrac{1}{2}\bar{u}\lambda,$$

where $\dot{m}$

is the rate of fluid mass hitting the surface,

ρ is the density of the fluid,

$\hat{u}$ is the average molecular speed $\bar{u} = \sqrt{\langle u^2 \rangle}$,

μ is the dynamic viscosity.

Effect of Temperature on the Viscosity of a Gas

Sutherland s formula can be used to derive the dynamic viscosity of an ideal gas as a function of the temperature:

$$\mu = \mu_0 \frac{T_0 + C}{T + C}\left(\frac{T}{T_0}\right)^{3/2}.$$

This in turn is equal to

$$\lambda \frac{T^{3/2}}{T + C}, \quad \text{where} \quad \lambda = \frac{\mu_0 (T_0 + C)}{T_0^{3/2}} \quad \text{is a constant.}$$

in Sutherland's formula:

- μ = dynamic viscosity in (Pa·s) at input temperature T,
- μ_0 = reference viscosity in (Pa·s) at reference temperature T_0,
- T = input temperature in kelvins,
- T_0 = reference temperature in kelvins,
- C = Sutherland's constant for the gaseous material in question.

Valid for temperatures between $0 < T < 555$ K with an error due to pressure less than 10% below 3.45 MPa.

Sutherland's constant and reference temperature for some gases

Gas	*C [K]*	*T_0 [K]*	*μ_0 [μPa s]*
air	120	291.15	18.27
nitrogen	111	300.55	17.81
oxygen	127	292.25	20.18
carbon dioxide	240	293.15	14.8
carbon monoxide	118	288.15	17.2
hydrogen	72	293.85	8.76
ammonia	370	293.15	9.82
sulfur dioxide	416	293.65	12.54
helium	79.4	273	19

Viscosity of a Dilute Gas

The Chapman-Enskog equation may be used to estimate viscosity for a dilute gas. This equation is based on a semi-theoretical assumption by Chapman and Enskog. The equation requires three empirically determined parameters: the collision diameter (σ), the maximum energy of attraction divided by the Boltzmann constant ($/\kappa$) and the collision integral ($\omega(T^*)$).

$$\mu_0 \times 10^6 = 2.6693 \frac{(MT)^{1/2}}{\sigma^2 \omega(T^*)},$$

with

- $T^* = \kappa T / \varepsilon$ — reduced temperature (dimensionless),
- μ_0 = viscosity for dilute gas (μPa.s),
- M = molecular mass (g/mol),
- T = temperature (K),

- σ = the collision diameter (Å),
- ε/k = the maximum energy of attraction divided by the Boltzmann constant (K),
- ω_{μ} = the collision integral.

Liquids

In liquids, the additional forces between molecules become important. This leads to an additional contribution to the shear stress though the exact mechanics of this are still controversial. Thus, in liquids:

- Viscosity is independent of pressure (except at very high pressure); and
- Viscosity tends to fall as temperature increases (for example, water viscosity goes from 1.79 cP to 0.28 cP in the temperature range from 0 °C to 100 °C); see temperature dependence of liquid viscosity for more details.

The dynamic viscosities of liquids are typically several orders of magnitude higher than dynamic viscosities of gases.

Viscosity of Blends of Liquids

The viscosity of the blend of two or more liquids can be estimated using the Refutas equation. The calculation is carried out in three steps.

The first step is to calculate the Viscosity Blending Number (VBN) (also called the Viscosity Blending Index) of each component of the blend:

$$\text{VBN} = 14.534 \times \ln\left[\ln(\text{v} + 0.8)\right] + 10.975$$

where v is the kinematic viscosity in centistokes (cSt). It is important that the kinematic viscosity of each component of the blend be obtained at the same temperature.

The next step is to calculate the VBN of the blend, using this equation:

$$\text{VBN}_{\text{Blend}} = \left[\text{x}_\text{A} \times \text{VBN}_\text{A}\right] + \left[\text{x}_\text{B} \times \text{VBN}_\text{B}\right] + \cdots + \left[\text{x}_\text{N} \times \text{VBN}_\text{N}\right]$$

where x_X is the mass fraction of each component of the blend.

Once the viscosity blending number of a blend has been calculated using equation (2), the final step is to determine the kinematic viscosity of the blend by solving equation (1) for v:

$$v = \exp\left(\exp\left(\frac{VBN_{Blend} - 10.975}{14.534}\right)\right) - 0.8,$$

where VBN_{Blend} is the viscosity blending number of the blend.

Viscosity of Selected Substances

The viscosity of air and water are by far the two most important materials for aviation aerodynamics and shipping fluid dynamics. Temperature plays the main role in determining viscosity.

Viscosity of Air

The viscosity of air depends mostly on the temperature. At 15.0 °C, the viscosity of air is 1.78×10^{-5} kg/(m·s), 17.8 μPa.s or 1.78×10^{-5} Pa.s.. One can get the viscosity of air as a function of temperature from the Gas Viscosity Calculator

Viscosity of Water

The dynamic viscosity of water is 8.90×10^{-4} Pa·s or 8.90×10^{-3} dyn·s/cm^2 or 0.890 cP at about 25 °C.

Water has a viscosity of 0.0091 poise at 25 °C, or 1 centipoise at 20 °C.

As a function of temperature T (K):

$$(\text{Pa·s}) = A \times 10^{B/(T-C)}$$

where A=2.414×10^{-5} Pa·s ; B = 247.8 K ; and C = 140 K.

Viscosity of liquid water at different temperatures up to the normal boiling point is listed below.

Temperature [°C]	*Viscosity [mPa—s]*
10	1.308
20	1.002
30	0.7978
40	0.6531
50	0.5471
60	0.4668
70	0.4044
80	0.3550
90	0.3150
100	0.2822

Viscosity of Various Materials

Viscosity of Slurry

The term slurry designs mixtures of a liquid and solid particles that retain some fluidity. The viscosity of slurry can be described as relative to the viscosity of the liquid phase:

$$\mu_s = \mu_r \cdot \mu_l,$$

where μ_s and μ_l are respectively the dynamic viscosity of the slurry and liquid (Pa·s), and μ_r is the relative viscosity (dimensionless).

Depending on the size and concentration of the solid particles, several models exist that describe the relative viscosity as a function of volume fraction x of solid particles.

In the case of extremely low concentrations of fine particles, Einstein's equation may be used:

$$\mu_r = 1 + 2.5 \cdot \phi$$

In the case of higher concentrations, a modified equation was proposed by Guth and Simha which takes into account interaction between the solid particles:

$$\mu_r = 1 + 2.5 \cdot \phi + 14.1 \cdot \phi^2$$

Further modification of this equation was proposed by Thomas from the fitting of empirical data:

$$\mu_r = 1 + 2.5 \cdot \phi + 10.05 \cdot \phi^2 + A \cdot e^{B \cdot \phi},$$

where A = *0.00273* and B = *16.6.*

In the case of very high concentrations, another empirical equation was proposed by Kitano *et al.* :

$$\mu_r = \left(1 - \frac{\phi}{A}\right)^{-2},$$

where A = *0.68* for smooth spherical particles.

Viscosity of Solids

On the basis that all solids such as granite flow in response to small shear stress, some researchers have contended that substances known as amorphous solids, such as glass and many polymers, may be considered to have viscosity. This has led some to the view that solids are simply "liquids" with a very high viscosity, typically greater than 10^{12} Pa·s. This position is often adopted by supporters of the widely held misconception that glass flow can be observed in old buildings. This distortion is the result of the undeveloped glass making process of earlier eras, and not due to the viscosity of glass.

However, others argue that solids are, in general, elastic for small stresses while fluids are not. Even if solids flow at higher stresses, they are characterized by their low-stress behaviour. This distinction is muddled if measurements are continued over long time periods, such as the Pitch drop experiment. Viscosity may be an appropriate characteristic for solids in a plastic regime. The situation becomes somewhat confused as the term *viscosity* is sometimes used for solid materials, for example Maxwell materials, to describe the relationship between stress and the rate of change of strain, rather than rate of shear.

These distinctions may be largely resolved by considering the constitutive equations of the material in question, which take into account both its viscous and elastic behaviors. Materials for which both their viscosity and their elasticity are important in a particular range of deformation and deformation rate are called *viscoelastic*. In geology, earth materials that exhibit viscous deformation at least three times greater than their elastic deformation are sometimes called rheids.

Viscosity of Amorphous Materials

Viscous flow in amorphous materials (e.g. in glasses and melts) is a thermally activated process:

$$\mu = A \cdot e^{Q/RT},$$

where Q is activation energy, T is temperature, R is the molar gas constant and A is approximately a constant.

The viscous flow in amorphous materials is characterized by a deviation from the Arrhenius-type behaviour: Q changes from a high value Q_H at low temperatures (in the glassy state) to a low value Q_L at high temperatures (in the liquid state). Depending on this change, amorphous materials are classified as either

- strong when: $Q_H - Q_L < Q_L$ or
- fragile when: $Q_H - Q_L \geq Q_L$.

The fragility of amorphous materials is numerically characterized by the Doremus' fragility ratio:

$$R_D = \frac{Q_H}{Q_L}$$

and strong material have $R_D < 2$ whereas fragile materials have $R_D \geq 2$.

The viscosity of amorphous materials is quite exactly described by a two-exponential equation:

$$\mu = A_1 \cdot T \cdot \left[1 + A_2 \cdot e^{B/RT}\right] \cdot \left[1 + C \cdot e^{D/RT}\right],$$

with constants A_1, A_2, B, C and D related to thermodynamic parameters of joining bonds of an amorphous material.

Not very far from the glass transition temperature, T_g, this equation can be approximated by a Vogel-Fulcher-Tammann (VFT) equation.

If the temperature is significantly lower than the glass transition temperature, $T < T_g$, then the two-exponential equation simplifies to an Arrhenius type equation:

$$\mu = A_L T \cdot e^{Q_H/RT}$$

with:

$$Q_H = H_d + H_m,$$

where H_d is the enthalpy of formation of broken bonds (termed configuron s) and H_m is the enthalpy of their motion. When the temperature is less than the glass transition temperature, $T < T_g$, the activation energy of viscosity is high because the amorphous materials are in the glassy state and most of their joining bonds are intact.

If the temperature is highly above the glass transition temperature, $T > T_g$, the two-exponential equation also simplifies to an Arrhenius type equation:

$$\mu = A_H T \cdot e^{Q_L/RT},$$

with:

$$Q_L = H_m.$$

When the temperature is higher than the glass transition temperature, $T > T_g$, the activation energy of viscosity is low because amorphous materials are melt and have most of their joining bonds broken which facilitates flow.

Eddy Viscosity

In the study of turbulence in fluids, a common practical strategy for calculation is to ignore the small-scale *vortices* (or *eddies*) in the motion and to calculate a large-scale motion with an *eddy viscosity* that characterizes the transport and dissipation of energy in the smaller-scale flow. Values of eddy viscosity used in modelling ocean circulation may be from 5×10^4 to 10^6 Pa·s depending upon the resolution of the numerical grid.

The Linear Viscous Stress Tensor

Viscous forces in a fluid are a function of the rate at which the fluid velocity is changing over distance. The velocity at any point *r* is specified by the velocity field *v*(*r*). The velocity at a small distance *dr* from point *r* may be written as a Taylor series:

$$\mathbf{v}(\mathbf{r}+\mathbf{dr}) = \mathbf{v}(\mathbf{r}) + \frac{\mathbf{dv}}{\mathbf{dr}}\mathbf{dr} + \ldots,$$

where *dv* / *dr* is shorthand for the dyadic product of the del operator and the velocity:

$$\frac{\mathbf{dv}}{\mathbf{dr}} = \begin{bmatrix} \frac{\partial v_x}{\partial x} & \frac{\partial v_x}{\partial y} & \frac{\partial v_x}{\partial z} \\ \frac{\partial v_y}{\partial x} & \frac{\partial v_y}{\partial y} & \frac{\partial v_y}{\partial z} \\ \frac{\partial v_z}{\partial x} & \frac{\partial v_z}{\partial y} & \frac{\partial v_z}{\partial z} \end{bmatrix}.$$

This is just the Jacobian of the velocity field.

Viscous forces are the result of relative motion between elements of the fluid, and so are expressible as a function of the velocity field. In other words, the forces at *r* are a function of *v*(*r*) and all derivatives of *v*(*r*) at that point. In the case of linear viscosity, the viscous force will be a function of the Jacobian tensor alone. For almost all practical situations, the linear approximation is sufficient.

If we represent *x*, *y*, and *z* by indices 1, 2, and 3 respectively, the *i,j* component of the Jacobian may be written as $\partial_i v_j$ where ∂_i is shorthand for $\partial/\partial x_i$. Note that when the first and higher derivative terms are zero, the velocity of all fluid elements is parallel, and there are no viscous forces.

Any matrix may be written as the sum of an antisymmetric matrix and a symmetric matrix, and this decomposition is independent of coordinate system, and so has physical significance. The velocity field may be approximated as:

$$v_j(\mathbf{r}+d\mathbf{r}) = v_j(\mathbf{r}) + \frac{1}{2}\left(\partial_i v_j - \partial_j v_i\right)dr_i + \frac{1}{2}\left(\partial_i v_j + \partial_j v_i\right)dr_i,$$

where Einstein notation is now being used in which repeated indices in a product are implicitly summed. The second term from the right is the asymmetric part of the first derivative term, and it represents a rigid rotation of the fluid about r with angular velocity ω where:

$$\omega = \frac{1}{2}\nabla\times\mathbf{v} = \frac{1}{2}\begin{bmatrix}\partial_2 v_3 - \partial_3 v_2\\ \partial_3 v_1 - \partial_1 v_3\\ \partial_1 v_2 - \partial_2 v_1\end{bmatrix}.$$

For such a rigid rotation, there is no change in the relative positions of the fluid elements, and so there is no viscous force associated with this term. The remaining symmetric term is responsible for the viscous forces in the fluid. Assuming the fluid is isotropic (i.e. its properties are the same in all directions), then the most general way that the symmetric term (the rate-of-strain tensor) can be broken down in a coordinate-independent (and therefore physically real) way is as the sum of a constant tensor (the rate-of-expansion tensor) and a traceless symmetric tensor (the rate-of-shear tensor):

$$\frac{1}{2}\left(\partial_i v_j + \partial_j v_i\right) = \underbrace{\frac{1}{3}\partial_k v_k \delta_{ij}}_{\text{rate-of-expansion tensor}} + \underbrace{\left(\frac{1}{2}\left(\partial_i v_j + \partial_j v_i\right) - \frac{1}{3}\partial_k v_k \delta_{ij}\right)}_{\text{rate-of-shear tensor}},$$

where δ_{ij} is the unit tensor. The most general linear relationship between the stress tensor σ and the rate-of-strain tensor is then a linear combination of these two tensors:

$$\sigma_{visc;ij} = \zeta\partial_k v_k \delta_{ij} + \mu\left(\partial_i v_j + \partial_j v_i - \frac{2}{3}\partial_k v_k \delta_{ij}\right),$$

where ς is the coefficient of bulk viscosity (or "second viscosity") and μ is the coefficient of (shear) viscosity.

The forces in the fluid are due to the velocities of the individual molecules. The velocity of a molecule may be thought of as the sum of the fluid velocity and the thermal velocity. The viscous stress tensor gives the force due to the fluid velocity only. The force on an area element in the fluid due to the thermal velocities of the molecules is just the hydrostatic pressure. This pressure term ($-p\,\delta_{ij}$) must be added to the viscous stress tensor to obtain the total stress tensor for the fluid.

The infinitesimal force dF_i on an infinitesimal area dA_i is then given by the usual relationship:

$$dF_i = \sigma_{ij}\, dA_j.$$

Creep (Deformation)

In materials science, creep is the tendency of a solid material to slowly move or deform permanently under the influence of stresses. It occurs as a result of long term exposure to high levels of stress that are below the yield strength of the material. Creep is more severe in materials that are subjected to heat for long periods, and near melting point. Creep always increases with temperature.

The rate of this deformation is a function of the material properties, exposure time, exposure temperature and the applied structural load. Depending on the magnitude of the applied stress and its duration, the deformation may become so large that a component can no longer perform its function — for example creep of a turbine blade will cause the blade to contact the casing, resulting in the failure of the blade. Creep is usually of concern to engineers and metallurgists when evaluating components that operate under high stresses or high temperatures. Creep is a deformation mechanism that may or may not constitute a failure mode. Moderate creep in concrete is sometimes welcomed because it relieves tensile stresses that might otherwise lead to cracking.

Unlike brittle fracture, creep deformation does not occur suddenly upon the application of stress. Instead, strain accumulates as a result of long-term stress. Creep is a "time-dependent" deformation.

The temperature range in which creep deformation may occur differs in various materials. For example, tungsten requires a temperature in the thousands of degrees before creep deformation can occur while ice will creep near 0 °C (32 °F). As a rule of thumb, the effects of creep deformation generally become noticeable at approximately 30% of the melting point (as measured on a thermodynamic temperature scale such as kelvin or rankine) for metals and 40–50% of melting point for ceramics. Virtually any material will creep upon approaching its melting temperature. Since the minimum temperature is relative to melting point, creep can be seen at relatively low temperatures for some materials. Plastics and low-melting-temperature metals, including many solders, creep at room temperature as can be seen markedly in old lead hot-water pipes. Glacier flow is an example of creep processes in ice.

Stages of Creep

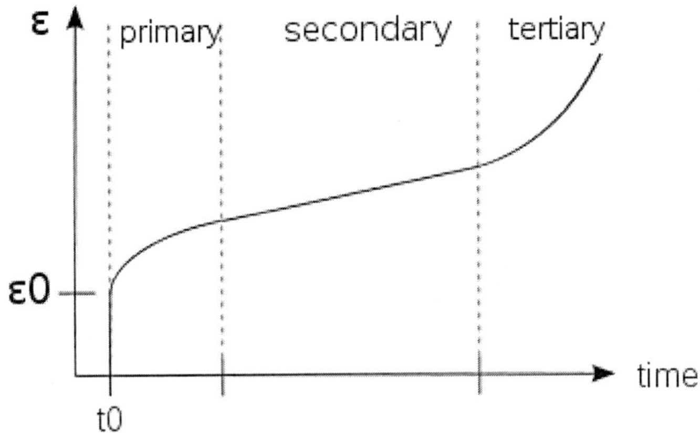

Figure 5: Strain as a function of time due to constant stress over an extended period for a viscoelastic material.

In the initial stage, or primary creep, the strain rate is relatively high, but slows with increasing time. This is due to work hardening. The strain rate eventually reaches a minimum and becomes near constant.

This is due to the balance between work hardening and annealing (thermal softening). This stage is known as secondary or steady-state creep. This stage is the most understood. The characterized "creep strain rate" typically refers to the rate in this secondary stage. Stress dependence of this rate depends on the creep mechanism. In tertiary creep, the strain rate exponentially increases with stress because of necking phenomena.

Mechanisms of Creep

The mechanism of creep depends on temperature and stress. The various methods are:

- Bulk diffusion (Nabarro-Herring creep)
- Climb — here the strain is actually accomplished by climb

- Climb-assisted glide — here the climb is an *enabling* mechanism, allowing dislocations to get around obstacles
- Grain boundary diffusion (Coble creep)
- Thermally activated glide — e.g., via cross-slip.

General Creep Equation

$$\frac{d\varepsilon}{dt} = \frac{C\sigma^m}{d^b} e^{\frac{-Q}{kT}}$$

where ε is the creep strain, *C* is a constant dependent on the material and the particular creep mechanism, *m* and *b* are exponents dependent on the creep mechanism, *Q* is the activation energy of the creep mechanism, σ is the applied stress, *d* is the grain size of the material, *k* is Boltzmann's constant, and *T* is the absolute temperature.

Dislocation Creep

At high stresses (relative to the shear modulus), creep is controlled by the movement of dislocations. For dislocation creep, Q = *Q(self diffusion)*, m = 4-6, and b = 0. Therefore, dislocation creep has a strong dependence on the applied stress and no grain size dependence.

Some alloys exhibit a very large stress exponent ($n > 10$), and this has typically been explained by introducing a "threshold stress," σ_{th}, below which creep can't be measured. The modified power law equation then becomes:

$$\frac{d\grave{o}}{dt} = A\left(\sigma - \sigma_{th}\right)^n e^{\frac{-Q}{\bar{R}T}}$$

where *A, Q* and *n* can all be explained by conventional mechanisms (so $3 \geq n \geq 10$).

Nabarro-Herring Creep

Nabarro-Herring creep is a form of diffusion controlled creep. In Nabarro-Herring creep, atoms diffuse through the lattice causing grains to elongate along the stress axis; *k* is related to the diffusion coefficient of atoms through the lattice, Q = *Q(self diffusion)*, m = 1, and b = 2. Therefore Nabarro-Herring creep has a weak stress dependence and a moderate grain size dependence, with the creep rate decreasing as grain size is increased.

Nabarro-Herring creep is strongly temperature dependent. For lattice diffusion of atoms to occur in a material, neighboring lattice sites or interstitial sites in the crystal structure must be free. A given atom must also overcome the energy barrier to move from its current site (it lies in an energetically favorable potential well) to the nearby vacant site (another potential well). The general form of the diffusion equation is $D = D_0 exp(E/KT)$ where D_0 has a dependence on both the attempted jump frequency and the number of nearest neighbour sites and the probability of the sites being vacant. Thus there is a double dependence upon temperature. At higher temperatures the diffusivity increases due to the direct temperature dependence of the equation, the increase in vacancies through Schottky defect formation, and an increase in the average energy of atoms in the material. Nabarro-Herring creep dominates at very high temperatures relative to a material's melting temperature.

Coble Creep

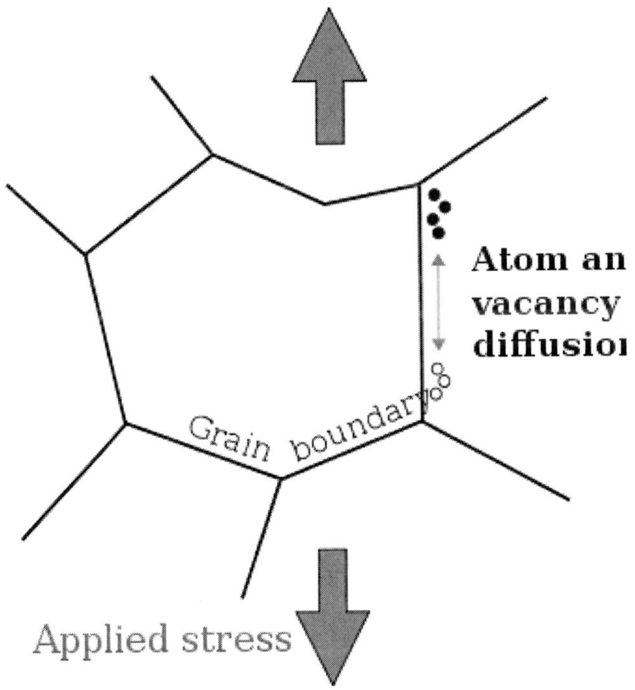

Figure 6: A Diagram Showing How Atoms and Vacancies Move Through a Grain as the Mechanism of Coble Creep

Coble creep, a form of diffusion creep, is a mechanism for deformation of crystalline solids. Coble creep occurs through the diffusion of atoms in a material along the grain boundaries, which produces a net flow of material and a sliding of the grain boundaries.

Coble creep is named after Robert L. Coble, who first reported his theory of how materials creep over time in 1962 in the Journal of Applied Physics.

The strain rate in a material experiencing Coble creep is given by:

$$\frac{\text{dò}}{\text{dt}} = \frac{\sigma}{d^3} D_{gb} e^{-Q_{Coble}/RT}$$

where

- σ is the applied stress
- d is the average grain boundary diameter
- D_{gb} is the diffusion coefficient in the grain boundary
- " Q_{Coble} is the activation energy for Coble creep
- R is the molar gas constant
- T is the temperature in kelvins

Note that in Coble creep, the strain rate $\frac{\text{dò}}{\text{dt}}$ is proportional to the applied stress σ; the same relationship is found for Nabarro-Herring creep. However, the two mechanisms differ in their relationship between the strain rate and grain size d.

In Coble creep, the strain rate is proportional to d^{-3}, whereas the strain rate in Nabarro-Herring creep is proportional to d^{-2}.

Researchers commonly use these relationships to determine which mechanism is dominant in a material; by varying the grain size and measuring how the strain rate is affected, they can determine the value of n in $\frac{\text{dò}}{\text{dt}} \alpha d^n$ and conclude whether Coble or Nabarro-Herring creep is dominant.

Creep of Polymers

Creep can occur in polymers and metals which are considered viscoelastic materials.

When a polymeric material is subjected to an abrupt force, the response can be modeled using the Kelvin-Voigt model. In this model, the material is represented by a Hookean spring and a Newtonian dashpot in parallel.

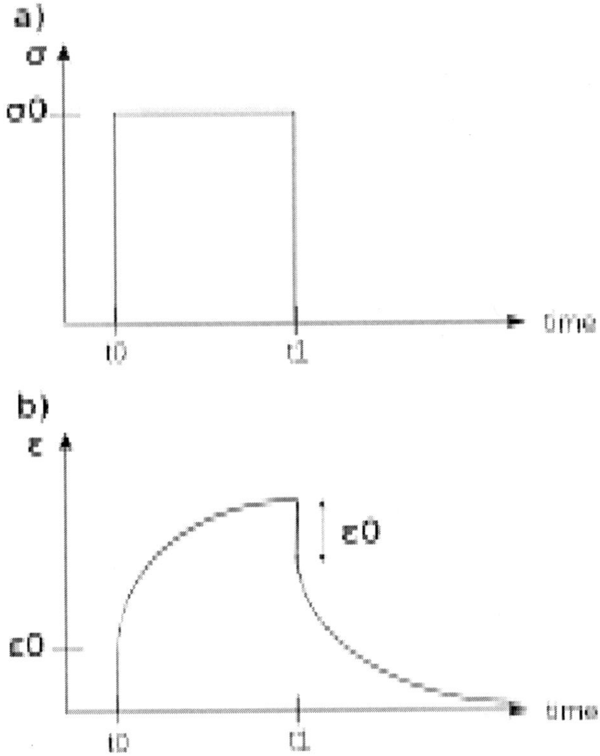

Figure 7: a) Applied Stress and b) Induced Strain as Functions of Time over a Short Period for a Viscoelastic Material.

The creep strain is given by

$$\varepsilon(t) = \sigma C_0 + \sigma C \int_0^\infty f(\tau)(1 - \exp[-t/\tau])d\tau$$

where:

- σ = applied stress
- C_0 = *instantaneous creep compliance*
- C = creep compliance coefficient
- τ = retardation time
- $f(\tau)$ = distribution of retardation times.

When subjected to a step constant stress, viscoelastic materials experience a time-dependent increase in strain. This phenomenon is known as viscoelastic creep. At a time t_0, a viscoelastic material is loaded with a constant stress that is maintained for a sufficiently long time period. The material responds to the stress with a strain that increases until the material ultimately fails. When the stress is maintained for a shorter time period, the material undergoes an initial strain until a time t_1 at which the stress is relieved, at which time the strain immediately decreases (discontinuity) then continues decreasing gradually to a residual strain.

Viscoelastic creep data can be presented in one of two ways. Total strain can be plotted as a function of time for a given temperature or temperatures. Below a critical value of applied stress, a material may exhibit linear viscoelasticity. Above this critical stress, the creep rate grows disproportionately faster. The second way of graphically presenting viscoelastic creep in a material is by plotting the creep modulus (constant applied stress divided by total strain at a particular time) as a function of time. Below its critical stress, the viscoelastic creep modulus is independent of stress applied. A family of curves describing strain versus time response to various applied stress may be represented by a single viscoelastic creep modulus versus time curve if the applied stresses are below the material's critical stress value.

Additionally, the molecular weight of the polymer of interest is known to affect its creep behaviour. The effect of increasing molecular weight tends to promote secondary bonding between polymer chains and thus make the polymer more creep resistant. Similarly, aromatic polymers are even more creep resistant due to the added stiffness from the rings. Both molecular weight and aromatic rings add to polymers' thermal stability, increasing the creep resistance of a polymer.

Both polymers and metals can creep. Polymers experience significant creep at temperatures above ca. –200°C; however, there are three main differences between polymeric and metallic creep.

Polymers show creep basically in two different ways. At typical work loads (5 up to 50%) ultra high molecular weight polyethylene (Spectra, Dyneema) will show time-linear creep, whereas polyester or aramids (Twaron, Kevlar) will show a time-logarithmic creep.

Applications

Though mostly due to the reduced yield stress at higher temperatures, the Collapse of the World Trade Centre was due in part to creep from increased temperature operation. The creep rate of hot pressure-loaded components in a nuclear reactor at power can be a significant design-constraint, since the creep rate is enhanced by the flux of energetic particles.

Creep was blamed for the Big Dig tunnel ceiling collapse in Boston, Massachusetts that occurred in July 2006.

An example of an application involving creep deformation is the design of tungsten light bulb filaments. Sagging of the filament coil between its supports increases with time due to creep deformation caused by the weight of the filament itself. If too much deformation occurs, the adjacent turns of the coil touch one another, causing an electrical short and local overheating, which quickly leads to failure of the filament. The coil geometry and supports are therefore designed to limit the stresses caused by the weight of the filament, and a special tungsten alloy with small amounts of oxygen trapped in the crystallite grain boundaries is used to slow the rate of Coble creep.

In steam turbine power plants, pipes carry steam at high temperatures (566 °C or 1050 °F) and pressures (above 24.1 MPa or 3500 psi). In jet engines, temperatures can reach up to 1400 °C (2550 °F) and initiate creep deformation in even advanced-coated turbine blades. Hence, it is crucial for correct functionality to understand the creep deformation behaviour of materials.

Creep deformation is important not only in systems where high temperatures are endured such as nuclear power plants, jet engines and heat exchangers, but also in the design of many everyday objects. For example, metal paper clips are stronger than plastic ones because plastics creep at room temperatures.

Aging glass windows are often erroneously used as an example of this phenomenon: measurable creep would only occur at temperatures above the glass transition temperature around 500 °C (900 °F). While glass does exhibit creep under the right conditions, apparent sagging in old windows may instead be a consequence of obsolete manufacturing processes, such as that used to create crown glass, which resulted in inconsistent thickness.

Melting

Melting, or fusion, is a physical process that results in the phase change of a substance from a solid to a liquid. The internal energy of a substance is increased, typically by the application of heat or pressure, resulting in a rise of its temperature to the melting point, at which the rigid ordering of molecular entities in the solid breaks down to a less-ordered state and the solid liquefies. An object that has melted completely is *molten*. Substances in the molten state generally have reduced viscosity with elevated temperature; an exception to this maxim is the element sulfur, whose viscosity increases with higher temperatures in its molten state.

Some organic compounds melt through mesophases, states of partial order between solid and liquid.

Melting Point

The melting point of a solid is the temperature at which it changes state from solid to liquid. At the melting point the solid and liquid phase exist in equilibrium. The melting point of a substance depends (usually slightly) on pressure and is usually specified at standard atmospheric pressure. When considered as the temperature of the reverse change from liquid to solid, it is referred to as the freezing point or crystallization point. Because of the ability of some substances to supercool, the freezing point is not considered as a characteristic property of a substance. When the “characteristic freezing point” of a substance is determined, in fact the actual methodology is almost always “the principle of observing the disappearance rather than the formation of ice”, that is, the melting point.

Examples

For most substances, melting and freezing points are approximately equal. For example, the melting point *and* freezing point of the element mercury is 234.32 kelvins (“38.83 °C or “37.89 °F). However, certain substances possess differing solid-liquid transition temperatures. For example, agar melts at 85 °C (185 °F) and solidifies from 31 °C to 40 °C (89.6 °F to 104 °F); this process is known as hysteresis.

The melting point of ice at 1 atmosphere of pressure is very close to 0 °C (32 °F, 273.15 K), this is also known as the ice point. In the presence of nucleating substances the freezing point of water is the same as the melting point, but in the absence of nucleators water can supercool to “42 °C (“43.6 °F, 231 K) before freezing.

The chemical element with the highest melting point is tungsten, at 3683 K (3410 °C, 6170 °F) making it excellent for use as filaments in light bulbs. The often-cited carbon does not melt at ambient pressure but sublimes at about 4000 K; a liquid phase only exists above pressures of 10 MPa and estimated 4300–4700 K. Tantalum hafnium carbide (Ta_4HfC_5) is a refractory compound with a very high melting point of 4488 K (4215 °C, 7619 °F). At the other end of the scale, helium does not freeze at all at normal pressure, even at temperatures very close to absolute zero; pressures over 20 times normal atmospheric pressure are necessary.

Melting Point Measurements

Many laboratory techniques exist for the determination of melting points. A Kofler bench is a metal strip with a temperature gradient (range from room temperature to 300 °C). Any substance can be placed on a section of the strip revealing its thermal behaviour at the temperature at that point. Differential scanning calorimetry gives information on melting point together with its enthalpy of fusion.

A basic melting point apparatus for the analysis of crystalline solids consists of a oil bath with a transparent window (most basic design: a Thiele tube) and a simple magnifier. The several grains of a solid are placed in a thin glass tube and partially immersed in the oil bath. The oil bath is heated (and stirred) and with the aid of the magnifier (and external light source) melting of the individual crystals at a certain temperature can be observed. In large/small devices, the sample is placed in a heating block, and optical detection is automated.

The measurement can also be made continuously with an operating process. For instance, oil refineries measure the freeze point of diesel fuel online, meaning that the sample is taken from the process and measured automatically. This allows for more frequent measurements as the sample does not have to be manually collected and taken to a remote laboratory.

Thermodynamics

Not only is heat required to raise the temperature of the solid to the melting point, but the melting itself requires heat called the heat of fusion.

From a thermodynamics point of view, at the melting point the change in Gibbs free energy (ΔG) of the material is zero, but the enthalpy (H) and the entropy (S) of the material are increasing (ΔH, $\Delta S > 0$</math>). Melting phenomenon happens when the Gibbs free energy of the liquid becomes lower than the solid for that material.

At various pressures this happens at a specific temperature. It can also be shown that:

$$\Delta S = \frac{\Delta H}{T}$$

Here *T*, *S* and *ΔH* are respectively the temperature at the melting point, change of entropy of melting and the change of enthalpy of melting.

The melting point is sensitive to extremely large changes in pressure, but generally this sensitivity is orders of magnitude less than the boiling point, because the solid-liquid transition represents only a small change in volume.

If, as observed in most cases, a substance is more dense in the solid than in the liquid state, the melting point will increase with increases in pressure. Otherwise the reverse behaviour occurs. Notably, this is the case of water, as illustrated graphically to the right, but also of Si, Ge, Ga, Bi. With extremely large changes in pressure, substantial changes to the melting point are observed.

For example, the melting point of silicon at ambient pressure (0.1 MPa) is 1415 °C, but at pressures in excess of 10 GPa it decreases to 1000 °C. Melting points are often used to characterize organic and inorganic compounds and to ascertain their purity. The melting point of a pure substance is always higher and has a smaller range than the melting point of an impure substance or, more generally, of mixtures.

The higher the quantity of other components, the lower the melting point and the broader will be the melting point range, often referred to as the pasty range. The temperature at which melting begins for a mixture is known as the solidus while the temperature where melting is complete is called the liquidus. Eutectics are special types of mixtures that behave like single phases. They melt sharply at a constant temperature to form a liquid of the same composition. Alternatively, on cooling a liquid with the eutectic composition will solidify as uniformly dispersed, small (fine-grained) mixed crystals with the same composition.

In contrast to crystalline solids, glasses do not possess a melting point; on heating they undergo a smooth glass transition into a viscous liquid. Upon further heating, they gradually soften, which can be characterized by certain softening points.

Freezing-point Depression

The freezing point of a solvent is depressed when another compound is added, meaning that a solution has a lower freezing point than a pure solvent. This phenomenon is used in technical applications to avoid freezing, for instance by adding salt or ethylene glycol to water.

Carnelley s Rule

In organic chemistry Carnelley's Rule, established in 1882 by Thomas Carnelley, stated that *high molecular symmetry is associated with high melting point*. Carnelley based his rule on examination of 15,000 chemical compounds. For example for three structural isomers with molecular formula C_5H_{12} the melting point increases in the series isopentane –160 °C (113 K) n-pentane –129.8 °C (143 K) and neopentane –18 °C (255 K). Likewise in xylenes and also dichlorobenzenes the melting point increases in the order meta, ortho and then para. Pyridine has a lower symmetry than benzene hence its lower melting point but the melting point again increases with diazine and triazines. Many cage-like compounds like adamantane and cubane with high symmetry have very high melting points.

A high melting point results from a high heat of fusion, a low entropy of fusion, or a combination of both. In highly symmetrical molecules the crystal phase is densely packed with many efficient intermolecular interactions resulting in a higher enthalpy change on melting.

Predicting the Melting Point of Substances (Lindemann s criterion)

An attempt to predict the bulk melting point of crystalline materials was first made in 1910 by Frederick Lindemann. The idea behind the theory was the observation that the average amplitude of thermal vibrations increase with increasing temperature. Melting initiates when the amplitude of vibration becomes large enough for adjacent atoms to partly occupy the same space. The Lindemann criterion states that melting is expected when the root mean square vibration amplitude exceeds a threshold value.

Assuming that all atoms in a crystal vibrate with the same frequency v, the average thermal energy can be estimated using the equipartition theorem as:

$$E = 4\pi^2 m v^2 u^2 = k_B T$$

where m is the atomic mass, v is the frequency, u is the average vibration amplitude, k_B is the Boltzmann constant, and T is the absolute temperature. If the threshold value of u^2 is c^2a^2 where c is the Lindemann constant and a is the atomic spacing, then the melting point is estimated as

$$T_m = \frac{4\pi^2 m v^2 c^2 a^2}{k_B}.$$

Several other expressions for the estimated melting temperature can be obtained depending on the estimate of the average thermal energy. Another commonly used expression for the Lindemann criterion is

$$T_m = \frac{m v^2 c^2 a^2}{k_B}.$$

From the expression for the Debye frequency for v, we have

$$T_m = \frac{2\pi m c^2 a^2 \theta_D^2 k_B}{h^2}$$

where θ_D is the Debye temperature and h is the Planck constant. Values of c range from 0.15–0.3 for most materials.

Open Melting Point Data

In February 2011 Alfa Aesar released over 10,000 melting points of compounds from their catalog as Open Data. This data has been curated and is freely available for download. This data has been used to create a random forest model for melting point prediction which is now available as a free to use webservice. Highly curated and open melting point data is also available from Nature Precedings

Melting Criteria

Among the theoretical criteria for melting, the Lindemann and Born criteria are those most frequently used as a basis to analyse the melting conditions. The Lindemann criterion states that melting occurs because of vibrational instability, e.g. crystals melt when the average amplitude of thermal vibrations of atoms is relatively high compared with interatomic distances, e.g. $<\delta u^2>1/2 > \delta_L R_s$, where δu is the atomic displacement, the Lindemann parameter δ_LH"0.20-0.25 and R_s is a half of the inter-atomic distance. The Lindemann melting criterion is supported by experimental data both for crystalline materials and for glass-liquid transitions in amorphous materials. The Born criterion is based on rigidity catastrophe caused by the vanishing elastic shear modulus, e.g. when the crystal no longer has sufficient rigidity to mechanically withstand load.

Supercooling

Supercooling, also known as undercooling, is the process of lowering the temperature of a liquid or a gas below its freezing point without it becoming a solid.

A liquid below its standard freezing point will crystallize in the presence of a seed crystal or nucleus around which a crystal structure can form. However, lacking any such nucleus, the liquid phase can be maintained all the way down to the temperature at which crystal homogeneous nucleation occurs. The homogeneous nucleation can occur above the glass transition where the system is an amorphous (non-crystalline) solid.

Pure water normally freezes at 273.15 K (0 °C or 32 °F) but it can also be "supercooled" at standard pressure down to its crystal homogeneous nucleation at almost 231 K ("42 °C/"43.6 °F). If, however, it is cooled at a rate on the order of 10^6 K/s, the crystal nucleation can be avoided and water becomes a glass. Its glass transition temperature is much colder and harder to determine, but studies estimate it at about 165 K (–108 °C/–162.4 °F). Glassy water can be heated up to approximately 150 K (–123 °C/–189.4 °F). In the range of temperatures between 231 K (–42 °C/–43.6 °F) and 150 K (–123 °C/ –189.4 °F) experiments find only crystal ice.

Droplets of supercooled water often exist in stratiform and cumulus clouds. Aircraft flying through these clouds seed an abrupt crystallization of these droplets, which can result in the formation of ice on the aircraft's wings or blockage of its instruments and probes, unless the aircraft are equipped with an appropriate de-icing system. Freezing rain is also caused by supercooled droplets.

The process opposite to supercooling, the melting of a solid above the freezing point, is much more difficult, and a solid will almost always melt at the same temperature for a given pressure. For this reason, it is the melting point which is usually identified, using melting point apparatus; even when the subject of a paper is "freezing-point determination", the actual methodology is "the principle of observing the disappearance rather than the formation of ice". It is, however, possible, at a given pressure to superheat a liquid above its boiling point without it becoming gaseous.

Supercooling is often confused with freezing-point depression. Supercooling is the cooling of a liquid below its freezing point without it becoming solid. Freezing point depression is when a solution can be cooled below the freezing point of the corresponding pure liquid due to the presence of the solute; an example of this is the freezing point depression that occurs when sodium chloride is added to pure water.

Constitutional Supercooling

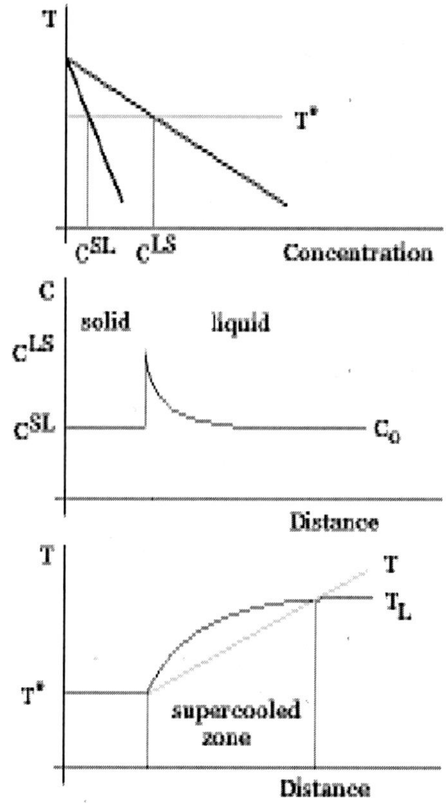

Figure 8: Constitutional supercooling – phase diagram, concentration, and temperature.

Constitutional supercooling occurs during solidification, is due to compositional changes, and results in cooling a liquid below the freezing point ahead of the solid–liquid interface. When solidifying a liquid, the interface is often unstable, and the velocity of the solid–liquid interface must be small in order to avoid constitutional supercooling.

Supercooled zones are observed when the liquidus temperature gradient at the interface is larger than the temperature gradient.

$$\left.\frac{\partial T_L}{\partial x}\right|_{x=0} > \frac{\partial T}{\partial x}$$

or

$$m\left.\frac{\partial C_L}{\partial x}\right|_{x=0} > \frac{\partial T}{\partial x}$$

The slope of the liquidus phase boundary on the phase diagram is $m = \partial T_L / \partial C_L$.

The concentration gradient is related to points, C^{LS} and C^{SL}, on the phase diagram:

$$\left.\frac{\partial C_L}{\partial x}\right|_{x=0} = -\frac{C^{LS} - C^{SL}}{D / v}$$

For steady-state growth $C^{SL} = C_0$ and the partition function $k = \frac{C^{SL}}{C^{LS}}$ can be assumed to be constant. Therefore the minimum thermal gradient necessary to create a stable solid front is as expressed below.

$$\frac{\partial T}{\partial x} < \frac{mC_0(1-k)v}{kD}$$

In Plants

Some plants are able to supercool the fluid in their cells cytosol and vacuole and thereby survive temperatures down to "40 °C. This is partly achieved through the synthesis of antifreeze proteins that prevent ice nucleation.

In Fish

The osmotic concentration of the body fluids of fish is lower that the osmotic concentration of sea water. Therefore the freezing point of fish can be above (!) the temperature of sea water. Of course, the freezing point can be lowered by anti-freeze agents. But there are fish, whose freezing point is indeed higher than the temperature of the surrounding sea water, and therefore the body fluids of these fish are supercooled. These fish must live well below the water surface, because they must not get in contact with ice (otherwise they would freeze immediately, since they are *only* supercooled.). Source: "Teleost fish have an osmotic concentration in their body fluids of about 300–400 mosm, and this corresponds to a freezing point of about “0.6 to “0.8 °C. Sea water in the polar regions often has a temperature of about “1,8 °C... Do they have a lower freezing point than ordinary fish, or do they remain supercooled throughout life? The answer is that both possibilities seem to have been realized.”, Knut Schmidt-Nielsen, *Animal Physiology, Adaption and Environment*. “In summer, the surface fish in the Hebron Fjord, Labrador, have no freezing problem. The fish that live deeper, however, where the water is at –1.73 °C, have a freezing point in their body fluids of –1.0 °C and must remain supercooled.

Applications

One commercial application of supercooling is in refrigeration. For example, there are freezers that cool drinks to a supercooled level so that when it is opened it slushes over. Another example is a product that can supercool the beverage in a conventional freezer. The Coca-Cola Company also briefly marketed special vending machines containing Sprite in the UK, and Coke in Singapore, which stored the bottles in a supercooled state so that they would turn to slush upon opening.

Melting of Amorphous Solids (Glasses)

Glasses are amorphous solids (e.g. amorphous materials that are at temperatures below the glass transition temperature) which are usually fabricated when the viscous molten material cools very rapidly to below its glass transition temperature, without sufficient time for a regular crystal lattice to form. Whether a material is liquid or solid depends primarily on the connectivity between its elementary building blocks so that solids are characterised by a high degree of connectivity whereas fluids occur at lower connectivity of the structural blocks. Melting of a solid material can also be considered as a percolation via broken connections between particles e.g. connecting bonds. In this approach melting of an amorphous material occurs when the broken bonds form a percolation cluster with T_g dependent on quasi-equilibrium thermodynamic parameters of bonds e.g. on enthalpy (H_d) and entropy (S_d) of formation of bonds in a given system at given conditions :

$$T_g=H_d/[S_d+R\ln[(1-f_c)/f_c]],$$

where f_c is the percolation threshold and R is the universal gas constant. Although H_d and S_d are not true equilibrium thermodynamic parameters and can depend on the cooling rate of a melt they can be found from available experimental data on viscosity of amorphous materials.

Premelting (Surface Melting)

Premelting (also: Surface melting) describes the fact that, even below its melting point T_s, quasi-liquid films can be observed on crystalline surfaces. The thickness of the film is temperature dependent. This effect is common for all crystalline materials. Premelting shows its effects in e.g. frost heave, the growth of snowflakes and, taking grain boundary interfaces into account, maybe even in the movement of glaciers.

Related Concepts

In genetics, melting DNA means to separate the double-stranded DNA into two single strands by heating or the use of chemical agents, cf. Polymerase chain reaction.

Electric Current

Electric current is a flow of electric charge through a medium. This charge is typically carried by moving electrons in a conductor such as wire. It can also be carried by ions in an electrolyte, or by both ions and electrons in a plasma.

The SI unit for measuring the rate of flow of electric charge is the ampere, which is charge flowing through some surface at the rate of one coulomb per second. Electric current is measured using an ammeter.

Symbol

The conventional symbol for current is I, which originates from the French phrase *intensité de courant*, or in English *current intensity*. This phrase is frequently used when discussing the value of an electric current, especially in older texts; modern practice often shortens this to simply *current* but *current intensity* is still used in many recent textbooks.

The I symbol was used by André-Marie Ampère, after whom the unit of electric current is named, in formulating the eponymous Ampère's force law which he discovered in 1820. The notation travelled from France to England, where it became standard, although at least one journal did not change from using C to I until 1896.

Conduction Mechanisms in Various Media

In metallic solids, electricity flows by means of electrons, from lower to higher electrical potential. In other media, any stream of charged objects may constitute an electric current. To provide a definition of current that is independent of the type of charge carriers flowing, *conventional current* is defined to flow in the same direction as positive charges. So in metals where the charge carriers (electrons) are negative, conventional current flows in the opposite direction as the electrons. In conductors where the charge carriers are positive, conventional current flows in the same direction as the charge carriers.

In a vacuum, a beam of ions or electrons may be formed. In other conductive materials, the electric current is due to the flow of both positively and negatively charged particles at the same time. In still others, the current is entirely due to positive charge flow. For example, the electric currents in electrolytes are flows of electrically charged atoms (ions), which exist in both positive and negative varieties.

In a common lead-acid electrochemical cell, electric currents are composed of positive hydrogen ions (protons) flowing in one direction, and negative sulfate ions flowing in the other. Electric currents in sparks or plasma are flows of electrons as well as positive and negative ions. In ice and in certain solid electrolytes, the electric current is entirely composed of flowing ions. In a semiconductor it is sometimes useful to think of the current as due to the flow of positive "holes" (the mobile positive charge carriers that are places where the semiconductor crystal is missing a valence electron). This is the case in a p-type semiconductor.

Metals

A solid conductive metal contains mobile, or free electrons, originating in the conduction electrons. These electrons are bound to the metal lattice but no longer to any individual atom. Even with no external electric field applied, these electrons move about randomly due to thermal energy but, on average, there is zero net current within the metal.

Given a surface through which a metal wire passes, the number of electrons moving from one side to the other in any period of time is on average equal to the number passing in the opposite direction. As George Gamow put in his science-popularizing book, *One, Two, Three...Infinity* (1947), "The metallic substances differ from all other materials by the fact that the outer shells of their atoms are bound rather loosely, and often let one of their electrons go free. Thus the interior of a metal is filled up with a large number of unattached electrons that travel aimlessly around like a crowd of displaced persons. When a metal wire is subjected to electric force applied on its opposite ends, these free electrons rush in the direction of the force, thus forming what we call an electric current."

When a metal wire is connected across the two terminals of a DC voltage source such as a battery, the source places an electric field across the conductor. The moment contact is made, the free electrons of the conductor are forced to drift toward the positive terminal under the influence of this field. The free electrons are therefore the charge carrier in a typical solid conductor. For an electric current of 1 ampere, 1 coulomb of electric charge (which consists of about 6.242×10^{18} elementary charges) drifts every second through any plane through which the conductor passes.

For a steady flow of charge through a surface, the current I (in amperes) can be calculated with the following equation:

$$I = \frac{Q}{t},$$

where Q is the electric charge transferred through the surface over some time t. If Q and t are measured in coulombs and seconds respectively, I is in amperes.

More generally, electric current can be represented as the rate at which charge flows through a given surface as:

$$I = \frac{dQ}{dt}.$$

Electrolytes

Electric currents in electrolytes are flows of electrically charged particles (ions). For example, if an electric field is placed across a solution of Na^+ and Cl^- (and conditions are right) the sodium ions move towards the negative electrode (cathode), while the chloride ions move towards the positive electrode (anode). Reactions take place at both electrode surfaces, absorbing each ion.

Water-ice and certain solid electrolytes called proton conductors contain positive hydrogen ions or "protons" which are mobile. In these materials, electric currents are composed of moving protons, as opposed to the moving electrons found in metals.

In certain electrolyte mixtures, brightly-coloured ions form the moving electric charges. The slow migration of these ions means that the current is visible.

Gases and Plasmas

In air and other ordinary gases below the breakdown field, the dominant source of electrical conduction is via a relatively small number of mobile ions produced by radioactive gases, ultraviolet light, or cosmic rays.

Since the electrical conductivity is low, gases are dielectrics or insulators. However, once the applied electric field approaches the breakdown value, free electrons become sufficiently accelerated by the electric field to create additional free electrons by colliding, and ionizing, neutral gas atoms or molecules in a process called avalanche breakdown.

The breakdown process forms a plasma that contains a significant number of mobile electrons and positive ions, causing it to behave as an electrical conductor. In the process, it forms a light emitting conductive path, such as a spark, arc or lightning.

Plasma is the state of matter where some of the electrons in a gas are stripped or "ionized" from their molecules or atoms. A plasma can be formed by high temperature, or by application of a high electric or alternating magnetic field as noted above. Due to their lower mass, the electrons in a plasma accelerate more quickly in response to an electric field than the heavier positive ions, and hence carry the bulk of the current.

Vacuum

Since a "perfect vacuum" contains no charged particles, it normally behaves as perfect insulator. However, metal electrode surfaces can cause a region of the vacuum to become conductive by injecting free electrons or ions through either field electron emission or thermionic emission. Thermionic emission occurs when the thermal energy exceeds the metal's work function, while field electron emission occurs when the electric field at the surface of the metal is high enough to cause tunnelling, which results in the ejection of free electrons from the metal into the vacuum.

Externally heated electrodes are often used to generate an electron cloud as in the filament or indirectly heated cathode of vacuum tubes. Cold electrodes can also spontaneously produce electron clouds via thermionic emission when small incandescent regions (called cathode spots or anode spots) are formed. These are incandescent regions of the electrode surface that are created by a localized high current flow. These regions may be initiated by field electron emission, but are then sustained by localized thermionic emission once a vacuum arc forms. These small electron-emitting regions can form quite rapidly, even explosively, on a metal surface subjected to a high electrical field. Vacuum tubes and sprytrons are some of the electronic switching and amplifying devices based on vacuum conductivity.

Current Density and Ohm s law

Current density is a measure of the density of an electric current. It is defined as a vector whose magnitude is the electric current per cross-sectional area. In SI units, the current density is measured in amperes per square meter.

$$I = \vec{J} \cdot \vec{A}$$

where I is current in the conductor, J is the current density, and A is the cross-sectional area. The dot product of the two vector quantities (A and J) is a scalar that represents the electric current.

Current density (current per unit area) J in a material is proportional to the conductivity σ and electric field E in the medium:

$$J = \sigma E$$

Instead of conductivity, reciprocal quantity called resistivity ρ, can be used:

$$J = \frac{E}{\rho}$$

Conduction in semiconductor devices may occur by a combination of electric field (drift) and diffusion, which is proportional to diffusion constant D and charge density α_q. The current density is then:

$$J = \sigma E + Dq\nabla n,$$

with q being the elementary charge and n the electron density. The carriers move in the direction of decreasing concentration, so for electrons a positive current results for a positive density gradient. If the carriers are holes, replace electron density n by the negative of the hole density p.

In linear anisotropic materials, σ, ρ and D are tensors.

In linear materials such as metals, and under low frequencies, the current density across the conductor surface is uniform.

In such conditions, Ohm's law states that the current is directly proportional to the potential difference between two ends (across) of that metal (ideal) resistor (or other ohmic device):

$$I = \frac{V}{R},$$

where I is the current, measured in amperes; V is the potential difference, measured in volts; and R is the resistance, measured in ohms.

For alternating currents, especially at higher frequencies, skin effect causes the current to spread unevenly across the conductor cross-section, with higher density near the surface, thus increasing the apparent resistance.

Drift Speed

The mobile charged particles within a conductor move constantly in random directions, like the particles of a gas. In order for there to be a net flow of charge, the particles must also move together with an average drift rate. Electrons are the charge carriers in metals and they follow an erratic path, bouncing from atom to atom, but generally drifting in the opposite direction of the electric field. The speed at which they drift can be calculated from the equation:

$$I = nAvQ,$$

where

I is the electric current

n is number of charged particles per unit volume (or charge carrier density)

A is the cross-sectional area of the conductor

v is the drift velocity, and

Q is the charge on each particle.

Electric currents in solids typically flow very slowly. For example, in a copper wire of cross-section 0.5 mm^2, carrying a current of 5 A, the *drift velocity* of the electrons is on the order of a millimetre per second. To take a different example, in the near-vacuum inside a cathode ray tube, the electrons travel in near-straight lines at about a tenth of the speed of light.

Any accelerating electric charge, and therefore any changing electric current, gives rise to an electromagnetic wave that propagates at very high speed outside the surface of the conductor. This speed is usually a significant fraction of the speed of light, as can be deduced from Maxwell's Equations, and is therefore many times faster than the drift velocity of the electrons.

For example, in AC power lines, the waves of electromagnetic energy propagate through the space between the wires, moving from a source to a distant load, even though the electrons in the wires only move back and forth over a tiny distance.

The ratio of the speed of the electromagnetic wave to the speed of light in free space is called the velocity factor, and depends on the electromagnetic properties of the conductor and the insulating materials surrounding it, and on their shape and size.

The magnitudes (but, not the natures) of these three velocities can be illustrated by an analogy with the three similar velocities associated with gases.

- The low drift velocity of charge carriers is analogous to air motion; in other words, winds.
- The high speed of electromagnetic waves is roughly analogous to the speed of sound in a gas (these waves move through the medium much faster than any individual particles do)
- The random motion of charges is analogous to heat- the thermal velocity of randomly vibrating gas particles.

This analogy is extremely simplistic and incomplete: The rapid propagation of a sound wave doesn't impart any change in the air molecules' drift velocity, whereas EM waves do carry the energy to propagate the actual current at a rate which is much, much higher than the electrons' drift velocity.

To illustrate the difference: The sound and the change in the air's drift velocity (the force of the wind gust) cross distance at rates equaling the speeds of sound and of mechanical transmission of force (not higher than rate of drift velocity); while a change in an EM field and the change in current (electrons' drift velocity) both propagate across distance at rates much higher than the actual drift velocity. You can hear wind much earlier than the force of the gust reaches you, but you don't observe a change in an EM field earlier than you can observe the change of current.

Electromagnetism

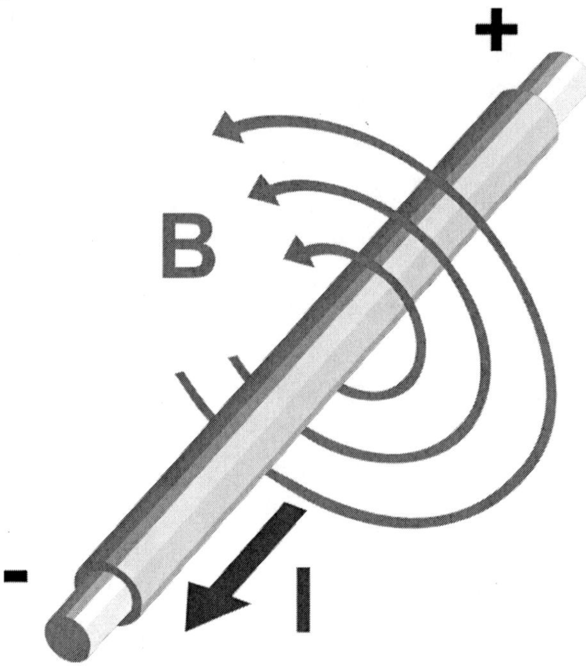

Figure 9: According to Ampère's law, an electric current produces a magnetic field.

Electric current produces a magnetic field. The magnetic field can be visualized as a pattern of circular field lines surrounding the wire.

Electric current can be directly measured with a galvanometer, but this method involves breaking the electrical circuit, which is sometimes inconvenient. Current can also be measured without breaking the circuit by detecting the magnetic field associated with the current. Devices used for this include Hall effect sensors, current clamps, current transformers, and Rogowski coils.

The theory of Special Relativity allows one to transform the magnetic field into a static electric field for an observer moving at the same speed as the charge in the diagram. The amount of current is particular to a reference frame.

Conventions

A flow of positive charges gives the same *electric* current, and has the same effect in a circuit, as an equal flow of negative charges in the opposite direction. Since current can be the flow of either positive or negative charges, or both, a convention for the direction of current which is independent of the type of charge carriers is needed. The direction of *conventional current* is defined arbitrarily to be the direction of the flow of positive charges.

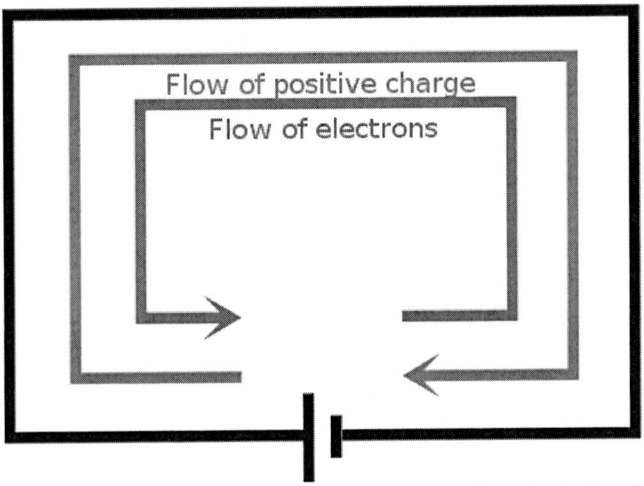

Figure 10: The electrons, the charge carriers in an electrical circuit, flow in the opposite direction of the conventional electric current.

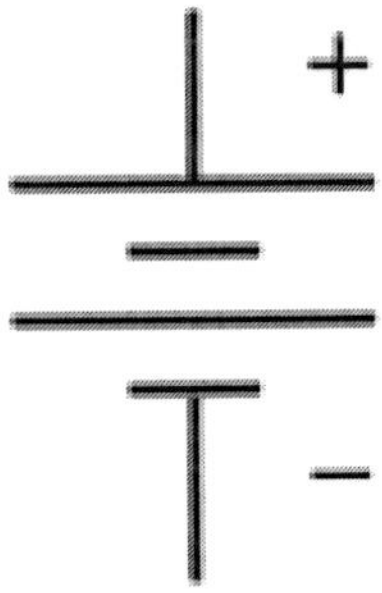

Figure 11: The symbol for a battery in a circuit diagram.

In metals, which make up the wires and other conductors in most electrical circuits, the positive charges are immobile, and only the negatively charged electrons flow. Because the electron carries negative charge, the *electron* motion in a metal conductor is in the direction opposite to that of conventional (or *electric*) current.

Reference Direction

When analysing electrical circuits, the actual direction of current through a specific circuit element is usually unknown. Consequently, each circuit element is assigned a current variable with an arbitrarily chosen *reference direction*. When the circuit is solved, the circuit element currents may have positive or negative values. A negative value means that the actual direction of current through that circuit element is opposite that of the chosen reference direction. In electronic circuits the reference current directions are usually chosen so that all currents flow toward ground. This often matches conventional current direction, because in many circuits the power supply voltage is positive with respect to ground.

Occurrences

Natural examples include lightning and the solar wind, the source of the polar auroras (the aurora borealis and aurora australis). The artificial form of electric current is the flow of conduction electrons in metal wires, such as the overhead power lines that deliver electrical energy across long distances and the smaller wires within electrical and electronic equipment. In electronics, other forms of electric current include the flow of electrons through resistors or through the vacuum in a vacuum tube, the flow of ions inside a battery or a neuron, and the flow of holes within a semiconductor.

Current Measurement

Current can be measured using an ammeter.

At the circuit level, there are various techniques that can be used to measure current:

- Shunt resistors
- Hall effect current sensor transducers
- Transformers (however DC cannot be measured)
- Magnetoresistive field sensors.

Mineralization (geology)

In geology, mineralization is the hydrothermal deposition of economically important metals in the formation of ore bodies or "lodes".

The first scientific studies of this process took place in Cornwall, United Kingdom by J.W.Henwood FRS and later by R.W. Fox, FRS.

The term can also refer to the process by which sediments replace organic material within the body of an organism that has died and was buried by sediments.

Mineralization may also refer to the product resulting from the process of mineralization. For example, mineralization (the process) may introduce metals (such as gold) into a rock. That rock may then be referred to as possessing gold mineralization.

Mineraloid

A mineraloid is a mineral-like substance that does not demonstrate crystallinity. Mineraloids possess chemical compositions that vary beyond the generally accepted ranges for specific minerals. For example, obsidian is an amorphous glass and not a crystal. Jet is derived from decaying wood under extreme pressure. Opal is another mineraloid because of its non-crystal nature. Pearls, considered by some to be a mineral because of the presence of calcium carbonate crystals within their structure, would be better considered a mineraloid because the crystals are bonded by an organic material and there is no definite proportion of the components.

Common Mineraloids

- Amber, non-crystalline structure
- Jet, non-crystalline nature, very compact coal
- Native mercury, liquid, IMA/CNMNC valid mineral name
- Lechatelierite, nearly pure silica glass
- Limonite, a mixture of oxides
- Lapis Lazuli, a mix of minerals
- Obsidian, volcanic glass- non-crystalline structure, a glass and quartz mixture
- Opal, non-crystalline silicon dioxide, a mix of minerals, IMA/ CNMNC valid mineral name
- Pearl, organicly produced carbonate
- Petroleum, liquid
- Pyrobitumen, non-homogeneous, non-crystalline structure, doesn't melt by heating
- Vulcanite, vulcanized natural or synthetic rubber, thus not a mineral due lack of crystalline structure
- Tektites, meteoritic silica glass.

Chapter 5

Mineral Processing

In the field of extractive metallurgy, mineral processing, also known as mineral dressing or ore dressing, is the process of separating commercially valuable minerals from their ores.

Figure 1 Crushing, a form of Comminution, one of the Unit Operations of Mineral Processing

History

Before the advent of heavy machinery the raw ore was broken up using hammers wielded by hand, a process called "spalling". Before long, mechanical means were found to achieve this. For instance, stamp mills were used in Samarkand as early as 973. They were also in use in medieval Persia. By the 11th century, stamp mills were in widespread use throughout the medieval Islamic world, from Islamic

Spain and North Africa in the west to Central Asia in the east. A later example was the Cornish stamps, consisting of a series of iron hammers mounted in a vertical frame, raised by cams on the shaft of a waterwheel and falling on to the ore under gravity.

The simplest method of separating ore from gangue consists of the picking out the individual crystals of each. This is a very tedious process, particularly when the individual particles are small. Another comparatively simple method relies on the various minerals having different densities, causing them to collect in different places: metallic minerals (being heavier) will drop out of suspension more quickly than lighter ones, which will be carried further by a stream of water. The process of panning and sifting for gold uses both of these methods. Various devices known as 'buddles' were used to take advantage of this property.[*when?*] Later, more advanced machines were used such as the Frue vanner, invented in 1874.

Other equipment used historically includes the hutch, a trough used with some ore-dressing machines and the keeve or kieve, a large tub used for differential settlement.

Unit Operations

Mineral processing can involve four general types of unit operation: *comminution* – particle size reduction; *sizing* – separation of particle sizes by screening or classification; *concentration* by taking advantage of physical and surface chemical properties; and *dewatering* – solid/ liquid separation. In all of these processes, the most important considerations are the economics of the processes and this is dictated by the grade and recovery of the final product. To do this, the mineralogy of the ore needs to be considered as this dictates the amount of liberation required and the processes that can occur. The smaller the particles processes, the greater the theortical grade and recovery of the final product, but this however is diffucult to do with fine particles as they prevent certain concentration processes from occurring.

Comminution

Comminution is particle size reduction of materials. Comminution may be carried out on either dry materials or slurries. Crushing and grinding are the two primary comminution processes. Crushing is normally carried out on "run-of-mine" ore, while grinding (normally carried out after crushing) may be conducted on dry or slurried material.

Sizing

Sizing is the general term for separation of particles according to their size.

The simplest sizing process is screening, or passing the particles to be sized through a screen or number of screens. Screening equipment can include grizzlies, bar screens, wedge wire screens, banna screens, multi-deck screens, vibratory screed, fine screens, flip flop screens and wire mesh screens. Screens can be static (typically the case for very coarse material), or they can incorporate mechanisms to shake or vibrate the screen. Some considerations in this process includes the screen material, the aperture size, shape and orientation, the amount of near sized particles, the addition of water, the amplitude and frequency of the vibrations, the angle of inclination, the presence of harmful materials, like steel and wood, and the size distribution of the particles. Classification refers to sizing operations that exploit the differences in settling velocities exhibited by particles of different size. Classification equipment may include ore sorters, gas cyclones, hydrocyclones, rotating trommels, rake classifiers or fluidized classifiers.

An important factor in both comminution and sizing operations is the determination of the particle size distribution of the materials being processed, commonly referred to as particle size analysis. Many techniques for analysing particle size are used, and the techniques include both off-line analyses which require that a sample of the material be taken for analysis and on-line techniques that allow for analysis of the material as it flows through the process.

Concentration

There are a number of ways to increase the concentration of the wanted minerals: in any particular case the method chosen will depend on the relative physical and surface chemical properties of the mineral and the gangue.

Gravity Concentration

Historically the earliest method used, particles can be classified based on their specific gravity. Air is the main fluid medium used for the process. Gravity concentration processes include:

- Heavy media or dense media separation (these include, baths, drums, larcodems, dyana whirlpool separators, and dense medium cyclones)

- Shaking tables, such as the Wilfley table
- Spiral separators
- Batac jigs
- Centrifugal bowl concentrators
- Jig concentrators are continuous processing gravity concentration devices using a pulsating fluidized bed.
- Multi gravity separators (Falcon, Knelson, Mozley and the Kelsey Jig)
- Inline pressure Jigs
- Reichert Cones.

These processes can be classified as either dense medium separation or gravity separation. The difference between the two that gravity separation does not use a dense medium to operate, only water or air. Dense medium separation can be performed with a variety of mediums. These include, organic liquids, aqueous solutions, suspensions in water and suspensions in air. Of these, most industrial processes use suspensions in water. The organic liquids are not used due to their toxicity and difficulties in handling. The aqueous solution as a dense medium is used in coal processing in the form of a belknap wash and the suspension in air is used in water-deficient areas, like china, where sand is used to separate coal from the gangue minerals. The dense medium separation is also classified as absolute gravity separation as the sinks and the floats travel in different directions. The gravity separation is also called relative gravity separation as they separate particles due to their differences in the magnitude of the particle response to a driving force.

These processes can also be classified into multi-G and single G processes. The difference is the magnitude of the driving force for the separation. Multi-G processes allow the separation of fine particles to occur and these particles can be in the range of 10 to 50 micron. The single G process are only capable of processing particles that are greater than 80 micron in diameter.

Of the gravity separation processes, the spiral concentrators are one of the most economical due to their simplicity and use of space. They operate by flowing film separation and can either use washwater or be washwater-less. The washwater spirals separate particles more easily but can have issues with entrainment of gangue with the concentrate produced..

Froth Flotation

Froth flotation is an important concentration process. This process can be used to separate any two different particles and operated by the surface chemistry of the particles. In flotation, bubbles are introduced into a pulp and the bubbles rise through the pulp. In the process, hydrophobic particles become bound to the surface of the bubbles.

The driving force for this attachment is the change in the surface free energy when the attachment occurs. These bubbles rise through the slurry and are collected from the surface. To enable these particles to attach, careful consideration of the chemistry of the pulp needs to be made. These considerations include the pH, Eh and the presence of flotation reagents. The pH is important as it changes the charge of the particles surface and the Eh effects the chemisorption of collectors on the surface of the particles.

The addition of flotation reagents also effects the operation of these processes. The most important chemical that is added is the collector, This chemical binds to the surface of the particles as it is a surfactant. The main considerations in this chemical is the nature of the head group and the size of the hydrocarbon chain. The hydrocarbon tail needs to be short to maximize the selectivity of the desired mineral and the headgroup dictates which minerals it attaches to. The frothers are another important chemical addition to the pulp at it enables stable bubbles to be formed. This is important as if the bubble coalesce, minerals fall off their surface. The bubbles however should not be too stable as this prevents easy transportation and dewatering of the concentrate formed. The mechanism of these frothers is not completely known and further research into their mechanisms is being performed.

Depressants and activators are used to selectively separate one mineral from another. Depressants inhibit the flotation of one mineral or minerals while activators enable the flotation of others. Examples of these include CN^-, used to depress all sulfides but galena and this depressant is believed to operate by changing the solubility of chemisorbed and physisorbed collectors on sulfides. This theory originates from Russia. An example of an activator is Cu^{2+} ions, used for the flotation of sphalerite.

There are a number of cells able to be used for the flotation of minerals. these include flotation columns and mechanical flotation cells. The flotation columns are used for finer minerals and they typically have a higher grade and lower recovery of minerals than mechanical flotation cells. The cells in use at the moment can exceed 300 m^3. This is done as they are cheaper per unit volume than smaller cells, but they are not able to be controlled as easily as smaller cells.

This process was invented in the 19th century in Australia. It was used to recover a sphalerite concentrate from tailings, produced using gravity concentration.

Further improvements have come from Australia in the form of the Jameson cell. This operated by the use of a plunging jet that generates fine bubbles. These fine bubbles have a higher kinetic energy and as such they can be used for the flotation of fine grained minerals, such as those produces by the isamill.

Electrostatic Separation

There are two main types of electrostatic separators. These work in similar ways, but the forces applied to the particles are different and these forces are gravity and electrostatic attraction. The two types are electrodynamic separators (or high tension rollers) or electrostatic separators. In high tension rollers, particles are charged by a corona discharge. This charges the particles that subsequently travel on a drum.

The conducting particles lose their charge to the drum and are removed from the drum with centripetal acceleration. Electrostatic plate separators work by passing a stream of particles past a charged anode. The conductors lose electrons to the plate and are pulled away from the other particles due to the induced attraction to the anode. These separators are used for particles between 75 and 250 micron and for efficient separation to occur, the particles need to be dry, have a close size distribution and uniform in shape. Of these considerations, one of the most important is the water content of the particles. This is important as a layer of moisture on the particles will render the non-conductors as conductors as the layer of the water is conductive.

Electrostatic plate separators are usually used for streams that have small conductors and coarse non-conductors. The high tension rollers are usually used for streams that have coarse conductors and fine non-conductors.

These separators are commonly used for separating mineral sands, an example of one of these mineral processing plants is the CRL processing plant at Pinkenba in Brisbane Queensland. In this plant, zircon, rutile and ilmenite are separated from the silica gangue. In this plant, the separation is performed in a number of stages with roughers, cleaners, scavengers and recleaners.

Magnetic Separation

Minerals such as ilmenite and magnetite are naturally magnetic, and so can be separated from non-magnetic particles using strong magnets. There are a number of different processes that can be used. These include high gradient magnetic separation (HGMS), high intensity magnetic separation (HIMS) and low intensity magnetic separation (LIMS). The HIMS and HGMS are differentiated as the HGMS separators are a batch process while the HIMS are a continuous process. These two processes are typically used for paramagnetic particles while the LIMS are used for ferromagnetic particles. The main considerations that need to be taken account when processing an ore is the size distribution, the presence of tramp metal and the liberation of the particles being separated.

This process operates by moving particles in a magnetic field. The force experienced in the magnetic field is given by the equation f=m/k.H.dh/dx. with k=magnetic susceptibility, H-magnetic field strength, and dh/dx being the magnetic field gradient. As seen in this equation, the separation can be driven in two ways, either through a gradient in a magnetic field or the strength of a magnetic field. The different driving forces are used in the different concentrators. These can be either with water or without. Like the spirals, washwater aids in the separation of the particles while increases the entrainment of the gangue in the concentrate.

Automated Ore Sorting

Modern, automated sorting applies optical sensors (visible spectrum, near infrared, X-ray, ultraviolet), that can be coupled with electrical conductivity and magnetic susceptibility sensors, to control the mechanical separation of ore into two or more categories on a individual rock by rock basis.

Dewatering

Dewatering is an important process in mineral processing. The purpose of dewatering is to remove water contained in particles. This is done for a number of reasons, specifically, to enable ore handling and concentrates to be transported easily, allow further processing to occur and to dispose of the gangue. The water removed from dewatering can be recycled through a mineral processing plant. The main processes that are used in dewatering include dewatering screens, sedimentation, filtering, and thermal drying. These processes increase in difficulty and cost as the particle size decreases.

Dewatering screens operate by passing particles over a screen. The particles pass over the screen while the water passes through the apertures in the screen. This process is only viable for coarse ores that have a close size distribution as the apertures can allow small particles to pass though and are not able to be produced for small particles.

Sedimentation operates by passing water into a large thickener or clarifier. In these devices, the particles settle out of the slurry under the effects of gravity or centripetal forces. These are limited by the surface chemistry of the particles and the size of the particles. To aid in the sedimentation process, floculants and coagulants are added to reduce the repulsive forces between the particles. This repulsive force is due to the double layer formed on the surface of the particles. The floculants work by binding multiple particles together while the coagulants work by reducing the thickness of the charged layer on the outside of the particle.

Thermal drying is usually used for fine particles and to remove low water content in the particles. Some common processes include rotary dryers, fluidised beds, spray driers, hearth dryers and rotary tray dryers. This process is usually expensive to operate due to the heating requirements of the dryers.

Other Processes

Many mechanical plants also incorporate hydrometallurgical or pyrometallurgical processes as part of an extractive metallurgical operation. Geometallurgy is a branch of extractive metallurgy that combines mineral processing with the geologic sciences.

A number of auxiliary materials handling operations are also considered a branch of mineral processing such as storage (as in bin design), conveying, sampling, weighing, slurry transport, and pneumatic transport.

Blast Furnace

A blast furnace is a type of metallurgical furnace used for smelting to produce industrial metals, generally iron.

In a blast furnace, fuel and ore and flux (limestone) are continuously supplied through the top of the furnace, while air (sometimes with oxygen enrichment) is blown into the bottom of the chamber, so that the chemical reactions take place throughout the furnace as the material moves downward. The end products are usually molten metal and slag phases tapped from the bottom, and flue gases exiting from the top of the furnace. The downward flow of the ore and flux in contact with an upflow of hot combustion gases is a countercurrent exchange process.

Blast furnaces are to be contrasted with air furnaces (such as reverberatory furnaces), which were naturally aspirated, usually by the convection of hot gases in a chimney flue. According to this broad definition, bloomeries for iron, blowing houses for tin, and smelt mills for lead would be classified as blast furnaces. However, the term has usually been limited to those used for smelting iron ore to produce pig iron, an intermediate material used in the production of commercial iron and steel.

History

Blast furnaces existed in China from about the 5th century BC, and in the West from the High Middle Ages. They spread from the region around Namur in Wallonia (Belgium) in the late 15th century, being introduced to England in 1491. The fuel used in these was invariably charcoal. The successful substitution of coke for charcoal is widely attributed to Abraham Darby in 1709. The efficiency of the process was further enhanced by the practice of preheating the blast, patented by James Beaumont Neilson in 1828.

The blast furnace is distinguished from the bloomery in that the object of the blast furnace is to produce molten metal that can be tapped from the furnace, whereas the intention in the bloomery is to avoid it melting so that carbon does not become dissolved in the iron. Bloomeries were also artificially blown using bellows, but the term *blast furnace* is normally reserved for furnaces where iron (or another metal) is refined from ore. Also, a blast furnace is countercurrent exchange process whereas a bloomery is not.

China

The oldest extant blast furnaces were built during the Han Dynasty of China in the 1st century BC. However, cast iron farm tools and weapons were widespread in China by the 5th century BC, while 3rd century BC iron smelters employed an average workforce of over two hundred men. These early furnaces had clay walls and used phosphorus-containing minerals as a flux. The effectiveness of the Chinese blast furnace was enhanced during this period by the engineer Du Shi (c. 31 AD), who applied the power of waterwheels to piston-bellows in forging cast iron.

While it was long thought that the Chinese had developed the blast furnace and cast iron as their first method of iron production, Donald Wagner has published a more recent paper that supersedes some of the statements in the earlier work; the newer paper still places the date of the first cast-iron artifacts at the 4th and 5th centuries BC, but also provides evidence of earlier bloomery furnace use, which migrated in from the West during the beginning of the Chinese Bronze Age of the late Longshan culture (2000 BC). He suggests that early blast furnace and cast iron production evolved from furnaces used to melt bronze. Certainly, though, iron was essential to military success by the time the State of Qin had unified China (221 BC). By the 11th century, the Song Dynasty Chinese iron industry made a remarkable switch of resources from charcoal to bituminous coal in casting iron and steel, sparing thousands of acres of woodland from felling. This may have happened as early as the 4th century AD.

The Chinese blast furnace remained in use well until the 20th century. The backyard furnaces favoured by Mao Zedong during the Great Leap Forward were of this type. In the regions with strong traditions of metallurgy, the steel production actually increased during this period. In the regions where there was no tradition of steelmaking or where the ironmasters knowing the traditional skills or the scientific principles of the blast furnace process had been killed, the results were less than satisfactory.

Ancient World Elsewhere

In most places in the world other than in China, there is no evidence of the use of the blast furnace (proper). Instead, iron was made by direct reduction in bloomeries. These are not correctly described as blast furnaces, though the term is occasionally misused in referring to them. An exception would the Haya people, who are renowned for creating steel using a blast furnace process and refining process very similar to open hearth process for possibly as long as 2000 years.

In Europe, the Greeks, Celts, Romans, and Carthaginians all used this process. Several examples have been found in France, and materials found in Tunisia suggest they were used there as well as in Antioch during the Hellenistic Period. Though little is known of it during the Dark Ages, the process probably continued in use. Similarly, smelting in bloomery-type furnaces in West Africa and forging for tools appear in the Nok culture in Africa by 500 BC. The earliest records of bloomery-type furnaces in East Africa are discoveries of smelted iron and carbon in Nubia and Axum that date back between 1,000-500 BCE. Particularly in Meroe, there are known to have been ancient bloomeries that produced metal tools for the Nubians and Kushites and produced surplus for their economy.

Medieval Europe

An improved bloomery, named the Catalan forge, was invented in Catalonia, Spain, during the 8th century. Instead of using natural draught, air was pumped in by bellows, resulting in better quality iron and an increased capacity. This pumping of airstream in with bellows is known as *cold blast*, and it increases the fuel efficiency of the bloomery and improves yield. The Catalan forges can also be built bigger than natural draught bloomeries.

Modern experimental archaeology and history re-enactment have shown there is only a very short step from the Catalan forge to the true blast furnace, where the iron is gained as pig iron in liquid phase. Usually, obtaining the iron in liquid phase is actually undesired, and the temperature is intentionally kept below the melting point of iron, since while removing the solid bloom mechanically is tedious and means batch process instead of continuous process, it is almost pure iron and can be worked immediately. On the other hand, pig iron is the eutectic mixture of carbon and iron and needs to be decarburized to produce steel or wrought iron, which was extremely tedious in the Middle Ages.

The oldest known blast furnaces in the West were built in Dürstel in Switzerland, the Märkische Sauerland in Germany, and at Lapphyttan in Sweden, where the complex was active between 1150 and 1350. At Noraskog in the Swedish county of Järnboås, there have also been found traces of blast furnaces dated even earlier, possibly to around 1100. These early blast furnaces, like the Chinese examples, were very inefficient compared to those used today. The iron from the Lapphyttan complex was used to produce balls of wrought iron known as osmonds, and these were traded internationally – a possible reference occurs in a treaty with Novgorod from 1203 and several certain references in accounts of English customs from the 1250s and 1320s. Other furnaces of the 13th to 15th centuries have been identified in Westphalia.

Knowledge of certain technological advances was transmitted as a result of the General Chapter of the Cistercian monks. This may have included the blast furnace, as the Cistercians are known to have been skilled metallurgists.

According to Jean Gimpel, their high level of industrial technology facilitated the diffusion of new techniques: "Every monastery had a model factory, often as large as the church and only several feet away, and waterpower drove the machinery of the various industries located on its floor." Iron ore deposits were often donated to the monks along with forges to extract the iron, and within time surpluses were being offered for sale. The Cistercians became the leading iron producers in Champagne, France, from the mid-13th century to the 17th century, also using the phosphate-rich slag from their furnaces as an agricultural fertilizer.

Archaeologists are still discovering the extent of Cistercian technology. At Laskill, an outstation of Rievaulx Abbey and the only medieval blast furnace so far identified in Britain, the slag produced was low in iron content. Slag from other furnaces of the time contained a substantial concentration of iron, whereas Laskill is believed to have produced cast iron quite efficiently. Its date is not yet clear, but it probably did not survive until Henry VIII's Dissolution of the Monasteries in the late 1530s, as an agreement (immediately after that) concerning the "smythes" with the Earl of Rutland in 1541 refers to blooms. Nevertheless, the means by which the blast furnace spread in medieval Europe has not finally been determined.

Early Modern Blast Furnaces: Origin and Spread

The direct ancestor of these used in France and England was in the Namur region in what is now Wallonia (Belgium). From there, they spread first to the Pays de Bray on the eastern boundary of Normandy and from there to the Weald of Sussex, where the first furnace (called Queenstock) in Buxted was built in about 1491, followed by one at Newbridge in Ashdown Forest in 1496. They remained few in number until about 1530 but many were built in the following decades in the Weald, where the iron industry perhaps reached its peak about 1590. Most of the pig iron from these furnaces was taken to finery forges for the production of bar iron.

The first British furnaces outside the Weald appeared during the 1550s, and many were built in the remainder of that century and the following ones. The output of the industry probably peaked about 1620, and was followed by a slow decline until the early 18th century. This was apparently because it was more economic to import iron from Sweden and elsewhere than to make it in some more remote British locations. Charcoal that was economically available to the industry was probably being consumed as fast as the wood to make it grew. The Backbarrow blast furnace built in Cumbria in 1711 has been described as the first efficient example.

The first blast furnace in Russia opened in 1637 near Tula and was called the Gorodishche Works. The blast furnace spread from here to the central Russia and then finally to the Urals.

The precursors of blast furnaces, bloomeries, have also been discovered and recorded to have been created in medieval West Africa with some of the metalworking Bantu civilizations such as the Bunyoro Empire and the Nyoro people.

Coke Blast Furnaces

In 1709, at Coalbrookdale in Shropshire, England, Abraham Darby began to fuel a blast furnace with coke instead of charcoal. Coke iron was initially only used for foundry work, making pots and other cast iron goods. Foundry work was a minor branch of the industry, but Darby's son built a new furnace at nearby Horsehay, and began to supply the owners of finery forges with coke pig iron for the production of bar iron. Coke pig iron was by this time cheaper to produce than charcoal pig iron. The use of a coal-derived fuel in the iron industry was a key factor in the British Industrial Revolution. Darby's old blast furnace has been archaeologically excavated and can be seen in situ at Coalbrookdale, part of the Ironbridge Gorge Museums. Cast iron from the furnace was used to make girders for the world's first iron bridge in 1779. The Iron Bridge crosses the River Severn at Coalbrookdale and remains in use for pedestrians.

A further important development was the change to hot blast, patented by James Beaumont Neilson at Wilsontown Ironworks in Scotland in 1828. This further reduced production costs. Within a few decades, the practice was to have a "stove" as large as the furnace next to it into which the waste gas (containing CO) from the furnace was directed and burnt. The resultant heat was used to preheat the air blown into the furnace.

A further significant development was the application of raw anthracite coal to the blast furnace, first tried successfully by George Crane at Ynyscedwyn ironworks in south Wales in 1837. It was taken up in America by the Lehigh Crane Iron Company at Catasauqua, Pennsylvania, in 1839.

Modern Furnaces

The blast furnace remains an important part of modern iron production. Modern furnaces are highly efficient, including Cowper stoves to pre-heat the blast air and employ recovery systems to extract the heat from the hot gases exiting the furnace. Competition in industry drives higher production rates. The largest blast furnaces have a volume around 5580 m^3 (190,000 cu ft) and can produce around 80,000 tonnes (88,000 short tons) of iron per week.

This is a great increase from the typical 18th-century furnaces, which averaged about 360 tonnes (400 short tons) per year. Variations of the blast furnace, such as the Swedish electric blast furnace, have been developed in countries which have no native coal resources.

Modern Process

Modern furnaces are equipped with an array of supporting facilities to increase efficiency, such as ore storage yards where barges are unloaded. The raw materials are transferred to the stockhouse complex by ore bridges, or rail hoppers and ore transfer cars. Rail-mounted scale cars or computer controlled weight hoppers weigh out the various raw materials to yield the desired hot metal and slag chemistry. The raw materials are brought to the top of the blast furnace via a skip car powered by winches or conveyor belts.

There are different ways in which the raw materials are charged into the blast furnace. Some blast furnaces use a "double bell" system where two "bells" are used to control the entry of raw material into the blast furnace. The purpose of the two bells is to minimize the loss of hot gases in the blast furnace. First, the raw materials are emptied into the upper or small bell. The bell is then rotated a predetermined amount to distribute the charge more accurately. The small bell then opens to empty the charge into the large bell. The small bell then closes, to seal the blast furnace, while the large bell dispenses the charge into the blast furnace. A more recent design is to use a "bell-less" system. These systems use multiple hoppers to contain each raw material, which is then discharged into the blast furnace through valves. These valves are more accurate at controlling how much of each constituent is added, as compared to the skip or conveyor system, thereby increasing the efficiency of the furnace. Some of these bell-less systems also implement a chute in order to precisely control where the charge is placed.

The iron making blast furnace itself is built in the form of a tall chimney-like structure lined with refractory brick. Coke, limestone flux, and iron ore (iron oxide) are charged into the top of the furnace in a precise filling order which helps control gas flow and the chemical reactions inside the furnace.

Four “uptakes” allow the hot, dirty gas to exit the furnace dome, while “bleeder valves” protect the top of the furnace from sudden gas pressure surges. When plugged, bleeder valves need to be cleaned with a bleeder cleaner.

The coarse particles in the gas settle in the “dust catcher” and are dumped into a railroad car or truck for disposal, while the gas itself flows through a venturi scrubber and a gas cooler to reduce the temperature of the cleaned gas.

The “casthouse” at the bottom half of the furnace contains the bustle pipe, tuyeres and the equipment for casting the liquid iron and slag. Once a “taphole” is drilled through the refractory clay plug, liquid iron and slag flow down a trough through a “skimmer” opening, separating the iron and slag. Modern, larger blast furnaces may have as many as four tapholes and two casthouses. Once the pig iron and slag has been tapped, the taphole is again plugged with refractory clay.

The tuyeres are used to implement a hot blast, which is used to increase the efficiency of the blast furnace. The hot blast is directed into the furnace through water-cooled copper nozzles called tuyeres near the base. The hot blast temperature can be from 900 °C to 1300 °C (1600 °F to 2300 °F) depending on the stove design and condition. The temperatures they deal with may be 2000 °C to 2300 °C (3600 °F to 4200 °F). Oil, tar, natural gas, powdered coal and oxygen can also be injected into the furnace at tuyere level to combine with the coke to release additional energy which is necessary to increase productivity.

Chemistry

The main chemical reaction producing the molten iron is:

$Fe_2O_3 + 3CO$ '& $2Fe + 3CO_2$

This reaction might be divided into multiple steps, with the first being that preheated blast air blown into the furnace reacts with the carbon in the form of coke to produce carbon monoxide and heat:

$$2\ C(s) + O_2(g) \rightarrow 2\ CO(g)$$

The hot carbon monoxide is the reducing agent for the iron ore and reacts with the iron oxide to produce molten iron and carbon dioxide. Depending on the temperature in the different parts of the furnace (warmest at the bottom) the iron is reduced in several steps. At the top, where the temperature usually is in the range between 200 °C and 700 °C, the iron oxide is partially reduced to iron(II,III) oxide, Fe_3O_4.

$$3\ Fe_2O_3(s) + CO(g) \rightarrow 2\ Fe_3O_4(s) + CO_2(g)$$

At temperatures around 850 °C, further down in the furnace, the iron(II,III) is reduced further to iron(II) oxide:

$$Fe_3O_4(s) + CO(g) \rightarrow 3\ FeO(s) + CO_2(g)$$

Hot carbon dioxide, unreacted carbon monoxide, and nitrogen from the air pass up through the furnace as fresh feed material travels down into the reaction zone. As the material travels downward, the counter-current gases both preheat the feed charge and decompose the limestone to calcium oxide and carbon dioxide:

$$CaCO_3(s) \rightarrow CaO(s) + CO_2(g)$$

As the iron(II) oxide moves down to the area with higher temperatures, ranging up to 1200 °C degrees, it is reduced further to iron metal:

$$FeO(s) + CO(g) \rightarrow Fe(s) + CO_2(g)$$

The carbon dioxide formed in this process is re-reduced to carbon monoxide by the coke:

$$C(s) + CO_2(g) \rightarrow 2\ CO(g)$$

The temperature-dependent equilibrium controlling the gas atmosphere in the furnace is called the Boudouard reaction:

$$2CO \rightleftharpoons CO_2 + C$$

The decomposition of limestone in the middle zones of the furnace proceeds according to the following reaction:

$$CaCO_3 \rightarrow CaO + CO_2$$

The calcium oxide formed by decomposition reacts with various acidic impurities in the iron (notably silica), to form a fayalitic slag which is essentially calcium silicate, $CaSiO_3$:

$$SiO_2 + CaO \rightarrow CaSiO_3$$

The "pig iron" produced by the blast furnace has a relatively high carbon content of around 4–5%, making it very brittle, and of limited immediate commercial use. Some pig iron is used to make cast iron. The majority of pig iron produced by blast furnaces undergoes further processing to reduce the carbon content and produce various grades of steel used for construction materials, automobiles, ships and machinery.

Although the efficiency of blast furnaces is constantly evolving, the chemical process inside the blast furnace remains the same. According to the American Iron and Steel Institute: "Blast furnaces will survive into the next millennium because the larger, efficient furnaces can produce hot metal at costs competitive with other iron making technologies." One of the biggest drawbacks of the blast furnaces is the inevitable carbon dioxide production as iron is reduced from iron oxides by carbon and there is no economical substitute – steelmaking is one of the unavoidable industrial contributors of the CO_2 emissions in the world.

The challenge set by the greenhouse gas emissions of the blast furnace is being addressed in an on-going European Programme called ULCOS (Ultra Low CO_2 Steelmaking). Several new process routes have been proposed and investigated in depth to cut specific emissions (CO_2 per ton of steel) by at least 50%. Some rely on the capture and further storage (CCS) of CO_2, while others choose decarbonizing iron and steel production, by turning to hydrogen, electricity and biomass. In the nearer term, a technology that incorporates CCS into the blast furnace process itself and is called the Top-Gas Recycling Blast Furnace is under development, with a scale-up to a commercial size blast furnace under way. The technology should be fully demonstrated by the end of the 2010s, in line with the timeline set, for example, by the EU to cut emissions significantly. Broad deployment could take place from 2020 on.

Manufacture of Stone Wool

Stone wool or rock wool is a spun mineral fibre used as an insulation product and in hydroponics. It is manufactured in a blast furnace fed with diabase rock which contains very low levels of metal oxides. The resultant slag is drawn off and spun to form the rock wool product. Very small amounts of metals are also produced which are an unwanted by-product and run to waste.

Decommissioned Blast Furnaces as Museum Sites

For a long time, it was normal procedure for a decommissioned blast furnace to be demolished and either be replaced with a newer, improved one, or to have the entire site demolished to make room for follow-up use of the area. In recent decades, several countries have realized the value of blast furnaces as a part of their industrial history. Rather than being demolished, abandoned steel mills were turned into museums or integrated into multi-purpose parks. The largest number of preserved historic blast furnaces exists in Germany; other such sites exist in Spain, France, the Czech Republic, Japan, Luxembourg, Poland, Romania, Mexico, Russia and the United States.

Roasting (metallurgy)

Roasting is a step in the processing of certain ores. More specifically, roasting is a metallurgical process involving gas–solid reactions at elevated temperatures with the goal of purifying the metal component(s). Often before roasting, the ore has already been partially purified, e.g. by froth floatation. The concentrate is mixed with other materials to facilitate the process. The technology is useful but is also a serious source of air pollution.

Roasting consists of thermal gas–solid reactions, which can include oxidation, reduction, chlorination, sulfation, and pyrohydrolysis. In roasting, the ore or ore concentrate is treated with very hot air. This process is generally applied to sulphide minerals. During roasting, the sulfide is converted to an oxide, and sulfur is released as sulfur dioxide, a gas. For the ores Cu_2S (chalcocite) and ZnS (sphalerite), balanced equations for the roasting are:

$$2\,Cu_2S + 3O_2 \rightarrow 2\,Cu_2O + 2\,SO_2$$

$$2\,ZnS + 3\,O_2 \rightarrow 2\,ZnO + 2\,SO_2$$

The gaseous product of sulfide roasting, sulfur dioxide (SO_2) is often used to produce sulfuric acid. Many sulfide minerals contain other components such as arsenic that are released into the environment.

Up until the early 20th century, roasting was started by burning wood on top of ore. This would raise the temperature of the ore to the point where its sulfur content would become its source of fuel, and the roasting process could continue without external fuel sources. Early sulfide roasting, was practiced in this manner, in "open hearth" roasters, which were manually stirred (a practice referred to as "rabbling") using rake-like tools to expose unroasted ore to oxygen as the reaction proceeded.

Multiple-hearth Roaster

In a multiple-hearth roaster, the concentrate drops through a series of 9 or more hearths stacked inside a brick-lined cylindrical column. As the feed concentrate drops through the furnace, it is first dried by the hot gases passing through the hearths and then oxidized to produce calcine. The reactions are slow and can be sustained only by the addition of fuel. Multiple hearth roasters are unpressurized and operate at about 690 °C (1,270 °F). Operating time depends upon the composition of concentrate and the amount of the sulfur removal required. Multiple hearth roasters have the capability of producing a high-purity calcine.

Suspension Roaster

In a suspension roaster, the concentrates are blown into a combustion chamber very similar to that of a pulverized coal furnace. The roaster consists of a refractory-lined cylindrical steel shell, with a large combustion space at the top and 2 to 4 hearths in the lower portion, similar to those of a multiple hearth furnace. Additional grinding, beyond that required for a multiple hearth furnace, is normally required to ensure that heat transfer to the material is sufficiently rapid for the desulfurization and oxidation reactions to occur in the furnace chamber. Suspension roasters are unpressurized and operate at about 980 °C (1,800 °F).

Fluidized-bed Roaster

In a fluidized-bed roaster, finely ground sulfide concentrates are suspended and oxidized in a feedstock bed supported on an air column. As in the suspension roaster, the reaction rates for desulfurization are more rapid than in the older multiple-hearth processes.

Fluidized-bed roasters operate under a pressure slightly lower than atmospheric and at temperatures averaging 1,000 °C (1,830 °F). In the fluidized-bed process, no additional fuel is required after ignition has been achieved. The major advantages of this roaster are greater throughput capacities, greater sulfur removal capabilities, and lower maintenance.

Electrolysis Process

The electrolysis process, also known as the hydrometallurgical process, Roast-Leach-Electrowin (RLE) process, or electrolytic process, is more widely used than the pyrometallurgical processes.

The electrolysis process consists of 4 steps: leaching, purification, electrolysis, and melting and casting.

Leaching (metallurgy)

Leaching is a widely used extractive metallurgy technique which converts metals into soluble salts in aqueous media. Compared to pyrometallurgical operations, leaching is easier to perform and much less harmful, because no gaseous pollution occurs. Drawbacks of leaching are the highly acidic and in some cases toxic residual effluent, and its lower efficiency caused by the low temperatures of the operation, which dramatically affect chemical reaction rates.

There are a variety of leaching processes, usually classified by the types of reagents used in the operation. The reagents required depend on the ores or pretreated material to be processed. A typical feed for leaching is either oxide or sulfide.

For material in oxide form, a simple acid leaching reaction can be illustrated by the zinc oxide leaching reaction :

$$ZnO + H_2SO_4 \rightarrow ZnSO_4 + H_2O$$

In this reaction solid ZnO dissolves, forming soluble zinc sulfate.

In many cases other reagents are used to leach oxides. For example, in the metallurgy of aluminium, aluminium oxide is subject to leaching by alkali solutions:

$$Al_2O_3 + 3H_2O + 2NaOH \rightarrow 2NaAl(OH)_4$$

Leaching of sulfides is a more complex process due to the refractory nature of sulfide ores. It often involves the use of pressurized vessels, called autoclaves. A good example of the autoclave leach process can be found in the metallurgy of zinc. It is best described by the following chemical reaction:

$$2ZnS + O_2 + 2H_2SO_4 \rightarrow 2ZnSO_4 + 2H_2O + 2S$$

This reaction proceeds at temperatures above the boiling point of water, thus creating a vapour pressure inside the vessel. Oxygen is injected under pressure, making the total pressure in the autoclave more than 0.6 MPa.

The leaching of precious metals such as gold can be carried out with cyanide or ozone under mild conditions.

Purification

The purification process utilizes the cementation process to further purify the zinc. It uses zinc dust and steam to remove copper, cadmium, cobalt, and nickel, which would interfere with the electrolysis process. After purification, concentrations of these impurities are limited to less than 0.05 milligram per liter (4×10^{-7} pound per U.S. gallon). Purification is usually conducted in large agitated tanks. The process takes place at temperatures ranging from 40 to 85 °C (104 to 185 °F), and pressures ranging from atmospheric to 2.4 atm (240 kPa) (absolute scale). The by-products are sold for further refining.

The zinc sulfate solution must be very pure for electrowinning to be at all efficient. Impurities can change the decomposition voltage enough to where the electrolysis cell produces largely hydrogen gas rather than zinc metal.

Electrolysis

Zinc is extracted from the purified zinc sulfate solution by electrowinning, which is a specialized form of electrolysis. The process works by passing an electric current through the solution in a series of cells. This causes the zinc to deposits on the cathodes (aluminum sheets) and oxygen to form at the anodes. Sulfuric acid is also formed in the process and reused in the leaching process. Every 24 to 48 hours, each cell is shut down, the zinc-coated cathodes are removed and rinsed, and the zinc is mechanically stripped from the aluminum plates.

Electrolytic zinc smelters contain as many as several hundred cells. A portion of the electrical energy is converted into heat, which increases the temperature of the electrolyte. Electrolytic cells operate at temperature ranges from 30 to 35 °C (86 to 95 °F) and at atmospheric pressure. A portion of the electrolyte is continuously circulated through the cooling towers both to cool and concentrate the electrolyte through evaporation of water. The cooled and concentrated electrolyte is then recycled to the cells. This process accounts for approximately one-third of all the energy usage when smelting zinc.

There are two common processes for electrowinning the metal: the *low current density* process, and the *Tainton high current density* process. The former uses a 10% sulfuric acid solution as the electrolyte, with current density of 270–325 amperes per square meter. The latter uses 22–28% sulfuric acid solution as the electrolyte with a current density of about 1,000 amperes per square meter. The latter gives better purity and has higher production capacity per volume of electrolyte, but has the disadvantage of running hotter and being more corrosive to the vessel in which it is done. In either of the electrolytic processes, each metric ton of zinc production expends about 3,900 kW·h (14 GJ) of electric power.

Melting and Casting

The final step is to melt the cathodes in an induction furnace. It is then either cast into pure zinc (99.995% pure) ingots or alloyed and cast into ingots.

Pyrometallurgical Processes

There are also several pyrometallurgical processes that reduce zinc oxide using carbon, then distill the metallic zinc from the resulting mix in an atmosphere of carbon monoxide. The major downfall of any of the pyrometallurgical process is that it is only 98% pure; a standard composition is 1.3% lead, 0.2% cadmium, 0.03% iron, and 98.5% zinc. This may be pure enough for galvanization, but not enough for die casting alloys, which requires special high-grade zinc (99.995% pure). In order to reach this purity the zinc must be refined.

The four types of commercial pyrometallurgical processes are the *Belgian-type horizontal retort* process, the *New Jersey Zinc continuous vertical-retort* process, the *blast furnace* process and the *St. Joseph Minerals Corporation s* (electrothermic) process.

St. Joseph Mineral Company (electrothermic) Process

This process was developed by the St. Joseph Mineral Company in 1930, and is the only pyrometallurgical process still used in the US to smelt zinc. The advantage of this system is that it is able to smelt a wide variety of zinc-bearing materials, including electric arc furnace dust. The disadvantage of this process is that it is less efficient than the electrolysis process.

The process begins with a downdraft sintering operation. The sinter, which is a mixture of roaster calcine and EAF calcine, is loaded onto a gate type conveyor and then combustions gasses are pumped through the sinter. The carbon in the combustion gases react with some the impurities, such as lead, cadmium, and halides. These impurities are driven off into filtration bags. The sinter after this process, called product sinter, usually has a composition of 48% zinc, 8% iron, 5% aluminum, 4% silicon, 2.5% calcium, and smaller quantities of magnesium, lead, and other metals. The sinter product is then charged with coke into an electric retort furnace. A pair of graphite electrodes from the top and bottom furnace produce current flow through the mixture. The coke provides electrical resistance to the mixture in order to heat the mixture to 1,400 °C (2,550 °F) and produce carbon monoxide. These conditions allow for the following chemical reaction to occur:

$$ZnO+CO \rightarrow Zn(vapor) + CO_2$$

The zinc vapour and carbon dioxide pass to a vacuum condenser, where zinc is recovered by bubbling through a molten zinc bath. Over 95% of the zinc vapour leaving the retort is condensed to liquid zinc. The carbon dioxide is regenerated with carbon, and the carbon monoxide is recycled back to the retort furnace.

Blast Furnace Process

This process was developed by the Imperial Smelting Corporation at Avonmouth, England, in order to increase production, increase efficiency, and decrease labour and maintenance costs. L. J. Derham proposed using a spray of molten lead droplets to rapidly chill and absorb the zinc vapour, despite the high concentration of carbon dioxide. The mixture is then cooled, where the zinc separates from the lead. The first plant using this design opened up in 1950. One of the advantages of this process is that it can co-produce lead bullion and copper dross. In 1990, it accounted for 12% of the world's zinc production.

The process starts by charging solid sinter and heated coke into the top of the blast furnace. Preheated air at 190 to 1,050 °C (370 to 1,920 °F) is blown into the bottom of the furnace. Zinc vapour and sulfides leave through the top and enter the condenser. Slag and lead collect at the bottom of the furnace and are tapped off regularly. The zinc is scrubbed from the vapour in the condenser via liquid lead. The liquid zinc is separated from the lead in the cooling circuit. Approximately 5,000 metric tons (5,500 short tons) of lead are required each year for this process, however this process recovers 25% more lead from the starting ores than other processes.

New Jersey Zinc Continuous Vertical Retort

The New Jersey Zinc process is no longer used to produce primary zinc in the U.S., Europe and Japan, however it still is used to treat secondary operations. This processes peaked in 1960, when 5% of the world production was done by this process. A modified version of this process is still used at Huludao, China, which produced 65,000 metric tons per year.

This process begins by roasting concentrates that are mixed with coal and briquetted in two stages. The briquettes are then heated in an autogenous coker at 700 °C (1,292 °F) and then charged into the retort. There are three reasons to briquette the calcine: to ensure free downward movement of the charge; to permit heat transfer across a practical size cross-section; to allow adequate porosity for the passage of reduced zinc vapour to the top of the retort. The reduced zinc vapour that is collected at the top of the retort is then condensed to a liquid.

Overpelt improved upon this design by using only one large condensation chamber, instead of many small ones, as it was originally designed. This allowed for the carbon monoxide to be recirculated into the furnaces for heating the retorts.

This process was licensed to the Imperial Smelting Company (ISC), in Avonmouth, England, which had a large vertical retort (VR) plant in production for many years. It was used until the mid 1970's when it was superseded by the company's Imperial Smelting Furnace (ISF) plant. The VR plant was demolished in 1975.

Belgian-type Horizontal Retort Process

This process was the main process used in Britain from the mid-19th century until 1951. The process was very inefficient as it was designed as a small scale batch operation. Each retort only produced 40 kilograms (88 lb) so companies would put them together in banks and used one large gas burner to heat all of them. The Belgian process requires redistillation to remove impurities of lead, cadmium, iron, copper, and arsenic.

History

Although some zinc artifacts appear to have been made during classical antiquity in Europe, the first production of zinc in quantity seems to have been in India and China. In India, zinc was produced at Zawar from the 12th to the 18th centuries. The sphalerite ore found here was presumably converted to zinc oxide via roasting, although no archaeological evidence of this has been found. Smelting is thought to have been done in sealed cylindrical clay retorts which were packed with a mixture of roasted ore, dolomite, and an organic material, perhaps cow dung, and then placed vertically in a furnace and heated to around 1100 °C. Carbon monoxide produced by the charring of the organic material would have reduced the zinc oxide to zinc vapour, which then liquefied in a conical clay condenser at the bottom of the retort, dripping down into a collection vessel. Over the period 1400–1800, production is estimated to have been about 200 kg/day. Zinc was also smelted in China from the mid-sixteenth century on.

Large-scale zinc production in Europe began with William Champion, who patented a zinc distillation process in 1738. In Champion's process, zinc ore (in this case, the carbonate, $ZnCO_3$) was sealed in large reduction pots with charcoal and heated in a furnace. The zinc vapour then descended through an iron condensing pipe until reaching a water-filled vessel at the bottom. Champion set up his first zinc works in Bristol, England, but soon expanded to Warmley and by 1754 had built four zinc furnaces there. Although Champion succeeded in producing about 200 tons of zinc, his business plans were not successful and he was bankrupt by 1769. However, zinc smelting continued in this area until 1880.

Early European zinc production also took place in Silesia, in Carinthia, and in Liège, Belgium. In the Carinthian process, used in works established in 1798 by Bergrath Dillinger, a wood-fuelled furnace heated a large number of small vertical retorts, and zinc vapour then dropped through a ceramic pipe into a common condensation chamber below. This process was out of use by 1840. The Belgian and Silesian processes both used horizontal retorts. In Silesia, Johann Ruhberg built a furnace to distill zinc in 1799, at first using pots but later changing to flat-bottomed retorts called "muffles", attached to horizontal tubes bent downwards in which the zinc condensed. The Silesian process eventually merged with the Belgian process. This process, developed by Jean-Jacques Daniel Dony, was introduced 1805–1810, and used retorts with a cylindrical cross-section. Condensers were horizontal clay tubes extending from the ends of the retorts. The merged "Belgo-Silesian" horizontal retort process was widely adopted in Europe by the third quarter of the 19th century, and later in the United States.

Experimental attempts to extract zinc via electrolysis begun in the 19th century, but the only commercially successful application before 1913 was a process, used in Great Britain and Austria, where zinc and chlorine were co-produced by electrolysis of an aqueous zinc chloride solution. The Anaconda Copper Company, at Anaconda, Montana, and the Consolidated Mining and Smelting Company, at Trail, British Columbia, both built successful electrolytic plants in 1915 using the currently used zinc sulfate process. This method has continued to grow in importance and in 1975 accounted for 68% of world zinc production.

The continuous vertical retort process was introduced in 1929 by the New Jersey Zinc Company. This process used a retort with silicon carbide walls, around 9 meters high and with a cross section of 2 by 0.3 meters. The walls of the retort were heated to 1300 °C and briquettes consisting of sintered zinc ore, coke, coal, and recycled material were fed into the top of the retort. Gaseous zinc was drawn off from the top of the column and, after a 20-hour journey through the retort, spent briquettes were removed from the bottom. To condense the gaseous zinc, the company first used a simple brick chamber with carborundum baffles, but efficiency was poor. During the 1940s a condenser was developed which condensed the zinc vapour on a spray of liquid zinc droplets, thrown up by an electrical impeller.

The electrothermic process, developed by the St. Joseph's Lead Company, was somewhat similar. The first commercial plant using this process was built in 1930 at the present site of Josephtown, Pennsylvania. The electrothermic furnace was a steel cylinder around 15 meters high and 2 meters in diameter, lined with firebrick. A mixture of sintered ore and coke was fed into the top of the furnace, and a current of 10,000–20,000 amperes, at a potential difference of 240 volts, was applied between carbon electrodes in the furnace, raising the temperature to 1200–1400 °C. An efficient condenser was devised for this process from 1931–1936; it consisted of a bath of liquid zinc which the exhaust gases were drawn through by suction. The zinc content of the gas stream was absorbed into the liquid bath.

The blast-furnace process was developed starting in 1943 at Avonmouth, England by the Imperial Smelting Corporation. It uses a spray of molten lead droplets to condense the zinc vapour.

Flodin Process

Flodin process is a process for manufacturing steel developed by Henning Flodin from Sweden.

Using a specially constructed electrical furnace a mixture of hematite and coal are smelted in a continuous process, with the reduced metal accumulating at the bottom of the furnace, where it can be tapped off.

Basic Oxygen Steelmaking

Basic oxygen steelmaking (BOS, BOP, BOF, and OSM), also known as Linz-Donawitz-Verfahren steelmaking or the oxygen converter process is a method of primary steelmaking in which carbon-rich molten pig iron is made into steel. Blowing oxygen through molten pig iron lowers the carbon content of the alloy and changes it into low-carbon steel. The process is known as *basic* due to the pH of the refractories—calcium oxide and magnesium oxide—that line the vessel to withstand the high temperature of molten metal.

The process was developed in 1948 by Robert Durrer and commercialized in 1952–1953 by Austrian VOEST and ÖAMG. The LD converter, named after the Austrian towns Linz and Donawitz (a district of Leoben) is a refined version of the Bessemer converter where blowing of air is replaced with blowing oxygen. It reduced capital cost of the plants, time of smelting, and increased labour productivity. Between 1920 and 2000, labour requirements in the industry decreased by a factor of 1,000, from more than 3 worker-hours per tonne to just 0.003. The vast majority of steel manufactured in the world is produced using the basic oxygen furnace; in 2000, it accounted for 60% of global steel output. Modern furnaces will take a charge of iron of up to 350 tons and convert it into steel in less than 40 minutes, compared to 10–12 hours in an open hearth furnace.

History

The basic oxygen process developed outside of traditional "big steel" environment. It was developed and refined by a single man, Swiss engineer Robert Durrer, and commercialized by two small steel companies in allied-occupied Austria, which had not yet recovered from the destruction of World War II.

In 1858, Henry Bessemer patented a steelmaking process involving oxygen blowing for decarburizing molten iron (UK Patent No. 2207). For nearly a hundred years commercial quantities of oxygen were not available at all or were too expensive, and the invention remained unused. During the World War II German (C. V. Schwartz), Belgian (John Miles) and Swiss (Durrer and Heinrich Heilbrugge) engineers proposed their versions of oxygen-blown steelmaking, but only Durrer and Heilbrugge brought it to mass-scale production.

In 1943, Durrer, formerly a professor at the Berlin Institute of Technology, returned to Switzerland and accepted a seat on the board of Roll AG, the country's largest steel mill. In 1947 he purchased the first small 2.5-ton experimental converter from the U. S., and on April 3, 1948 the new converter produced its first steel. The new process could conveniently process large amounts of scrap metal with only a small proportion of primary metal necessary. In the summer of 1948 Roll AG and two Austrian state-owned companies, VOEST and ÖAMG, agreed to commercialize the Durrer process.

By June 1949, VOEST developed an adaptation of Durrer's process, the LD (Linz-Donawitz) process. In December 1949, the VOEST and the ÖAMG committed to building their first 30-ton oxygen converters. They were put into operation in November 1952 (VOEST in Linz) and May 1953 (ÖAMG, Donawitz) and temporarily became *the* leading edge of the world's steelmaking, causing a surge in steel-related research. 34 thousand businesspeople and engineers visited the VOEST converter by 1963. The LD process reduced processing time and capital costs per ton of steel, contributing to the competitive advantage of Austrian steel. VOEST eventually acquired the rights to market the new technology. However, errors made by the VOEST and the ÖAMG management in licensing their technology made control over its adoption in Japan impossible and by the end of the 1950s the Austrians lost their competitive edge.

The original LD process employed blowing oxygen on top of molten iron through a water-cooled nozzle of a vertical lance. In the 1960s steelmakers introduced bottom-blown converters and introduced inert gas blowing for stirring the molten metal and removing of phosphorus impurities.

The big American steelmakers did not catch up to the new technology; the first oxygen converters in the United States were launched in the end of 1954 by McLouth Steel in Trenton, Michigan, which had less than 1 per cent of the national steel market. U.S. Steel and Bethlehem Steel introduced oxygen process only in 1964. Elsewhere, owing to the Austrian licensing blunder, the technology was adopted quickly. By 1970 half of world's and 80% of Japan's steel output was produced in oxygen converters. In the last quarter of the 20th century basic oxygen converters were gradually substituted with the electric arc furnaces. In Japan the share of LD process decreased from 80% in 1970 to 70% in 2000; worldwide share of the basic oxygen process stabilized at 60%.

Process

The basic oxygen steel-making process is as follows:

1. Molten pig iron (sometimes referred to as "hot metal") from a blast furnace is poured into a large refractory-lined container called a ladle;
2. The metal in the ladle is sent directly for basic oxygen steelmaking or to a pretreatment stage. Pretreatment of the blast furnace metal is used to reduce the refining load of sulfur, silicon, and phosphorus. In desulfurising pretreatment, a lance is lowered into the molten iron in the ladle and several hundred kilograms of powdered magnesium are added. Sulfur impurities are reduced to magnesium sulfide in a violent exothermic reaction. The sulfide is then raked off. Similar pretreatment is possible for desiliconisation and dephosphorisation using mill scale (iron oxide) and lime as reagents. The decision to pretreat depends on the quality of the blast furnace metal and the required final quality of the BOS steel.
3. Filling the furnace with the ingredients is called *charging*. The BOS process is autogenous: the required thermal energy is produced during the process. Maintaining the proper *charge balance*, the ratio of hotmetal to scrap, is therefore very important. The BOS vessel is one-fifth filled with steel scrap. Molten iron from the ladle is added as required by the charge balance. A typical chemistry of hotmetal charged into the BOS vessel is: 4% C, 0.2–0.8% Si, 0.08%–0.18% P, and 0.01–0.04% S.
4. The vessel is then set upright and a water-cooled lance is lowered down into it. The lance blows 99% pure oxygen onto the steel and iron, igniting the carbon dissolved in the steel and burning it to form carbon monoxide and carbon dioxide, causing the temperature to rise to about 1700°C. This melts the scrap, lowers the carbon content of the molten iron and helps remove unwanted chemical elements. It is this use of oxygen instead of air that improves upon the Bessemer process, for the nitrogen (and other gases) in air do not react with the charge as oxygen does. High purity oxygen is blown into the furnace or BOS vessel through a vertically oriented water-cooled lance with velocities faster than Mach 1.

5. Fluxes (burnt lime or dolomite) are fed into the vessel to form slag, which absorbs impurities of the steelmaking process. During blowing the metal in the vessel forms an emulsion with the slag, facilitating the refining process. Near the end of the blowing cycle, which takes about 20 minutes, the temperature is measured and samples are taken. The samples are tested and a computer analysis of the steel given within six minutes. A typical chemistry of the blown metal is 0.3–0.6% C, 0.05–0.1% Mn, 0.01–0.03% Si, 0.01–0.03% S and P.
6. The BOS vessel is tilted again and the steel is poured into a giant ladle. This process is called *tapping* the steel. The steel is further refined in the ladle furnace, by adding alloying materials to give the steel special properties required by the customer. Sometimes argon or nitrogen gas is bubbled into the ladle to make sure the alloys mix correctly. The steel now contains 0.1–1% carbon. The more carbon in the steel, the harder it is, but it is also more brittle and less flexible.
7. After the steel is removed from the BOS vessel, the slag, filled with impurities, is poured off and cooled.

Chapter 6

Corrosion

Corrosion is the disintegration of an engineered material into its constituent atoms due to chemical reactions with its surroundings. In the most common use of the word, this means electrochemical oxidation of metals in reaction with an oxidant such as oxygen. Formation of an oxide of iron due to oxidation of the iron atoms in solid solution is a well-known example of electrochemical corrosion, commonly known as rusting. This type of damage typically produces oxide(s) and/or salt(s) of the original metal. Corrosion can also occur in materials other than metals, such as ceramics or polymers, although in this context, the term degradation is more common.

In other words, corrosion is the wearing away of metals due to a chemical reaction.

Many structural alloys corrode merely from exposure to moisture in the air, but the process can be strongly affected by exposure to certain substances. Corrosion can be concentrated locally to form a pit or crack, or it can extend across a wide area more or less uniformly corroding the surface. Because corrosion is a diffusion controlled process, it occurs on exposed surfaces.

As a result, methods to reduce the activity of the exposed surface, such as passivation and chromate-conversion, can increase a material's corrosion resistance. However, some corrosion mechanisms are less visible and less predictable.

Galvanic Corrosion

Galvanic corrosion is an electrochemical process in which one metal corrodes preferentially to another when both metals are in electrical contact and immersed in an electrolyte. The same galvanic reaction is exploited in primary batteries to generate a voltage.

Overview

Dissimilar metals and alloys have different electrode potentials and when two or more come into contact in an electrolyte a galvanic couple is set up. The potenetial difference between the dissimilar metals is the driving force for the accelerated attack on the anode member of the galvanic couple.

The electrolyte provides a means for ion migration whereby metallic ions can move from the anode to the cathode. This leads to the anodic metal corroding more quickly than it otherwise would; the corrosion of the cathodic metal is retarded even to the point of stopping. The presence of an electrolyte and a electronic conducting path between the metals is essential for galvanic corrosion to occur.

In some cases, this reaction is intentionally encouraged. For example, low-cost household batteries typically contain carbon-zinc cells. As part of a closed circuit (the electron pathway), the zinc within the cell will corrode preferentially (the ion pathway). Another example is the cathodic protection of buried or submerged structures. In this example, sacrificial anodes work as part of a galvanic couple, promoting corrosion of the anode, rather than the protected subject metal.

In other cases, such as HVAC systems with bimetallic plumbing/ piping (most commonly copper and cast iron) galvanic corrosion will contribute to accelerated corrosion rates of the system. Corrosion inhibitors such as sodium nitrite or sodium molybdate can be introduced to these systems to reduce the galvanic potential. However, the application of these corrosion inhibitors must be monitored closely.

If the application of corrosion inhibitors increases the conductivity of the water within the system, the galvanic corrosion potential can be exponentially increased. pH is also a major consideration with regards to closed loop bimetallic circulating systems. Should the pH and corrosion inhibition doses be incorrect, galvanic corrosion will be accelerated. In most HVAC systems, the concept of anodes and cathodes are not an option as they would need to be applied within the plumbing of the system and, over time, would corrode and cause potential mechanical damage to circulating pumps, heat exchangers, etc.

Galvanic Series

Metals (and their alloys) can be arranged in a galvanic series representing the potential they develop in a given electrolyte against a standard reference electrode. The relative position of two metals on such a series gives a good indication of which metal is more likely to corrode more quickly. However, other factors such as water aeration and flow rate can influence the process markedly.

Galvanic corrosion is of major interest to the marine industry. Galvanic series tables for seawater are commonplace due to the extensive use of metal in shipbuilding. It is possible that corrosion of silver brazing in a salt water pipe caused a failure that led to the *USS Thresher* sinking with all men lost. The common technique of cleaning silver by immersion of the silver and a piece of aluminium in a salt water bath (usually sodium bicarbonate) is an example of galvanic corrosion. (Care should be exercised for reasons such as this will strip silver oxide from the silver which may be there for decoration. Use on plated silver is inadvisable as this may introduce unwanted galvanic corrosion with the base metal.)

Preventing Galvanic Corrosion

There are several ways of reducing and preventing this form of corrosion.

- One way is to electrically insulate the two metals from each other. Unless they are in electrical contact, there can be no galvanic couple set up. This can be done using plastic or another insulator to separate steel water pipes from copper-based fittings or by using a coat of grease to separate aluminium and steel parts. Use of absorbent washers that may retain fluid is often counter-productive. Piping can be isolated with a spool of pipe made of plastic materials or made of metal material internally coated or lined. It is important that the spool has a minimum length of approx 500 mm to be effective.
- Another way is to keep the metals dry and/or shielded from ionic compounds (salts, acids, bases), for example by painting or encasing the protected metal in plastic or epoxy, and allowing them to dry.
- Coating the two materials or if it is not possible to coat both, the coating shall be applied to the more noble, the material with higher potential. This is necessary because if the coating is applied only on the more active material, in case of damage of the coating there will be a large cathode area and a very small anode area, and for the area effect the corrosion rate will be very high.

- It is also possible to choose metals that have similar potentials. The more closely matched the individual potentials, the lesser the potential difference and hence the lesser the galvanic current. Using the same metal for all construction is the most precise way of matching potentials.

Electroplating or other plating can also help. This tends to use more noble metals that resist corrosion better. Chrome, nickel, silver and gold can all be used.

Cathodic protection uses one or more sacrificial anodes made of a metal which is more active than the protected metal. Metals commonly used for sacrificial anodes include zinc, magnesium, and aluminium. This is commonplace in water heaters. Failure to regularly replace sacrificial anodes in water heaters severely diminishes the lifetime of the tank.

Finally, an electrical power supply may be connected to oppose the corrosive galvanic current.

For example, consider a system is composed of 316 SS (a 300 series stainless steel; it is a very noble alloy meaning it is quite resistant to corrosion and has a high potential) and a mild steel (a very active metal with lower potential).

The mild steel will corrode in the presence of an electrolyte such as salt water. If a sacrificial anode is used (such as a zinc alloy, aluminium alloy, or magnesium), these anodes will corrode, protecting the other metals.

This is a common practice in the marine industry to protect ship equipment. Boats and vessels that are in salt water use either zinc alloy or aluminium alloy. If boats are only in fresh water, a magnesium alloy is used. Magnesium has one of the highest galvanic potentials of any metal. If it is used in a salt water application on a steel or aluminium hull boat, hydrogen bubbles will form under the paint, causing blistering and peeling.

Metal boats connected to a mains shore line will normally have to have the hull connected to earth for safety reasons. However the end of that earth connection is likely to be a copper rod buried within the marina, resulting in a steel-copper "battery" of about 0.5V. For such cases the use of a galvanic isolator is essential- typically 2 diodes in series, preventing any current flow while the applied voltage is less than 1.4V (i.e. 0.7V per diode), but allowing a full flow in case of an earth fault. It has been noted that there will still be a very minor leak through the diodes which may result in slightly faster corrosion than normal.

Examples

A common example of galvanic corrosion is the rusting of corrugated iron sheet, which becomes widespread when the protective zinc coating is broken and the underlying steel is attacked. The zinc is attacked preferentially because it is less noble, but when consumed, rusting will occur in earnest. With a tin can, the opposite is true because the tin is more noble than the underlying steel, so when the coating is broken, the steel is attacked preferentially.

Statue of Liberty

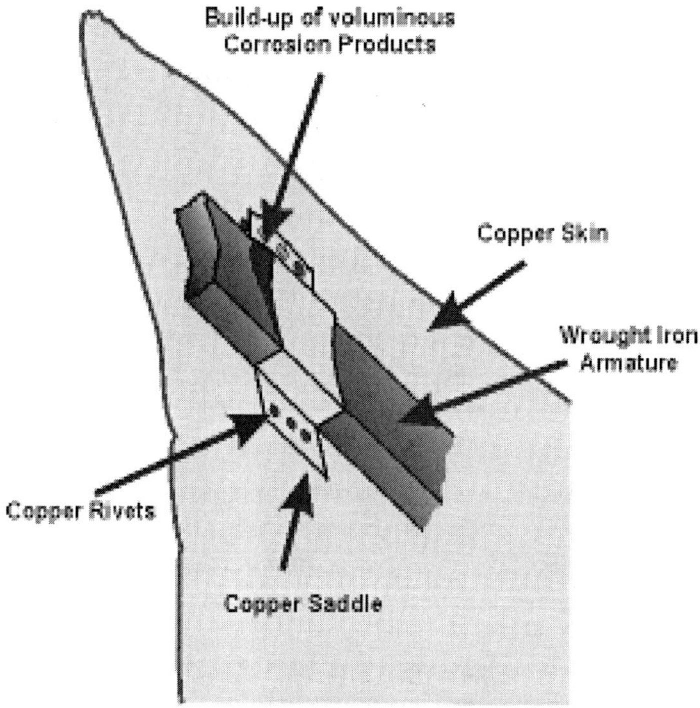

Figure 1 : Galvanic Corrosion in Statue of Liberty.

A rather more spectacular example occurred in the Statue of Liberty when regular maintenance in the 1980s showed that galvanic corrosion had taken place between the outer copper skin and the wrought iron support structure.

Although the problem had been anticipated when the structure was built by Gustave Eiffel to Frédéric Bartholdi's design in the 1880s, the insulation of shellac between the two metals failed over a period of time and resulted in rusting of the iron supports. The renovation replaced the original insulation with PTFE. The structure was far from unsafe owing to the large number of unaffected connections, but it was regarded as a precautionary measure for what is considered a national symbol of the United States.

Royal Navy and HMS Alarm

In 17th-century England, Samuel Pepys then serving as Admiralty Secretary, agreed to the removal of lead sheathing on British Royal Navy vessels to prevent the mysterious disintegration of their rudder-irons and bolt-heads—though he confessed himself baffled as to the reason the lead caused the corrosion.

The problem recurred when vessels were sheathed in copper. The protective copper on the Royal Navy frigate HMS *Alarm* detached from the wooden hull in many places because the iron nails which had been used to fasten the copper to the timbers had been "much rotted." Closer inspection revealed that some nails, which were less corroded, were insulated from the copper by brown paper which was trapped under the nail head.

The copper had been delivered to the dockyard wrapped in the paper which was not removed before the sheets were nailed to the hull. The obvious conclusion therefore, and the one which was contained in a report to the Admiralty of 1763, was that iron should not be allowed direct contact with copper in sea water to avoid severe corrosion.

Later ships were designed with this in mind. Not only was sea water a very good electrolyte owing to its high salt concentration, but attack of the nails was encouraged by their very small exposed area compared with that of the copper-sheathed hull.

Lasagna Cell

A "lasagna cell" is accidentally produced when salty food such as lasagna is stored in a steel baking pan and is covered with aluminium foil. After a few hours the foil develops small holes where it touches the lasagna, and the food surface becomes covered with small spots composed of corroded aluminium.

In this example, the salty food (e.g. lasagna) is the electrolyte, the aluminium foil the anode and the steel pan the cathode. If the aluminium foil only touches the electrolyte in small areas, the galvanic corrosion is concentrated and corrosion can occur fairly rapidly.

Galvanic Compatibility

Often when design requires that dissimilar metals come in contact, the galvanic compatibility is managed by finishes and plating. The finishing and plating selected facilitate the dissimilar materials being in contact and protect the base materials from corrosion.

Harsh environments, such as outdoors, high humidity, and salt environments, fall into this category. Typically there should be not more than 0.15 V difference in the "Anodic Index". For example; gold-silver would have a difference of 0.15V being acceptable.

For normal environments, such as storage in warehouses or non-temperature and humidity controlled environments, there should not be more than 0.25 V difference in the "Anodic Index". For controlled environments, in which temperature and humidity are controlled, 0.50 V can be tolerated.

Table 1 : Anodic index

Metal	*Index (V)*
Most Cathodic	
Gold, solid and plated, Gold-platinum alloy	0.00
Rhodium plated on silver-plated copper	0.05
Silver, solid or plated; monel metal. High nickel-copper alloys	0.15
Nickel, solid or plated, titanium an s alloys, Monel	0.30
Copper, solid or plated; low brasses or bronzes; silver solder; German silvery high copper-nickel alloys; nickel-chromium alloys	0.35
Brass and bronzes	0.40
High brasses and bronzes	0.45
18% chromium type corrosion-resistant steels	0.50
Chromium plated; tin plated; 12% chromium type corrosion-resistant steels	0.60
Tin-plate; tin-lead solder	0.65
Lead, solid or plated; high lead alloys	0.70
2000 series wrought aluminum	0.75
Iron, wrought, gray or malleable, plain carbon and low alloy steels	0.85
Aluminum, wrought alloys other than 2000 series aluminum, cast alloys of the silicon type	0.90

Aluminum, cast alloys other than silicon type, cadmium, plated and chromate	0.95
Hot-dip-zinc plate; galvanized steel	1.20
Zinc, wrought; zinc-base die-casting alloys; zinc plated	1.25
Magnesium & magnesium-base alloys, cast or wrought	1.75
Beryllium	1.85
Most Anodic	

Galvanic Series

The galvanic series (or electropotential series) determines the nobility of metals and semi-metals. When two metals are submerged in an electrolyte, while electrically connected, the less noble (base) will experience galvanic corrosion. The rate of corrosion is determined by the electrolyte and the difference in nobility. The difference can be measured as a difference in voltage potential. Galvanic reaction is the principle upon which batteries are based.

Galvanic series (most noble at top)

The following is the galvanic series for stagnant (that is, low oxygen content) seawater. The order may change in different environments.

- Graphite
- Palladium
- Platinum
- Gold
- Silver
- Titanium
- Stainless steel 316 (passive)
- Stainless Steel 304 (passive)
- Silicon bronze
- Stainless Steel 316 (active)
- Monel 400
- Phosphor bronze
- Admiralty brass
- Cupronickel
- Molybdenum

- Red brass
- Brass plating
- Yellow brass
- Naval brass 464
- Uranium 8% Mo
- Niobium 1% Zr
- Tungsten
- Stainless Steel 304 (active)
- Tantalum
- Chromium plating
- Nickel (passive)
- Copper
- Nickel (active)
- Cast iron
- Steel
- Lead
- Tin
- Indium
- Aluminum
- Uranium (pure)
- Cadmium
- Beryllium
- Zinc plating
- Magnesium.

Corrosion Removal

Often it is possible to chemically remove the products of corrosion to give a clean surface, but one that may exhibit artifacts of corrosion such as pitting. For example phosphoric acid in the form of naval jelly is often applied to ferrous tools or surfaces to remove rust.

Corrosion removal should not be confused with Electropolishing which removes some layers of the underlying metal to make a smooth surface. For example phosphoric acid (again) may be used to electropolish copper but it does this by removing copper, not the products of copper corrosion. We can also remove some part of corrosion by polishing the corroded metal surface by sandpaper and thereafter by applying oil completely to the part of polishing and then wash the metal surface completely and dry that part.

After that paint the metal surface to protect it from further corrosion. And its costless also to follow the process and maintain the removal without highly electroplating the surface.

Resistance to Corrosion

Some metals are more intrinsically resistant to corrosion than others, either due to the fundamental nature of the electrochemical processes involved or due to the details of how reaction products form. If a more susceptible material is used, many techniques can be applied during an item's manufacture and use to protect its materials from damage.

Intrinsic Chemistry

The materials most resistant to corrosion are those for which corrosion is thermodynamically unfavorable. Any corrosion products of gold or platinum tend to decompose spontaneously into pure metal, which is why these elements can be found in metallic form on Earth, and is a large part of their intrinsic value. More common "base" metals can only be protected by more temporary means.

Some metals have naturally slow reaction kinetics, even though their corrosion is thermodynamically favorable. These include such metals as zinc, magnesium, and cadmium. While corrosion of these metals is continuous and ongoing, it happens at an acceptably slow rate. An extreme example is graphite, which releases large amounts of energy upon oxidation, but has such slow kinetics that it is effectively immune to electrochemical corrosion under normal conditions.

Passivation

Passivation is the process of making a material "passive", usually by the deposition of a layer of oxide that adheres to the metal surface. In air, passivation affects the properties of almost all metals–notable examples being aluminium, zinc, titanium, and silicon (a metalloid). In the context of corrosion, passivation is the spontaneous formation of a hard non-reactive surface film that inhibits further corrosion. This layer is usually an oxide or nitride that is a few nanometers thick.

Mechanisms

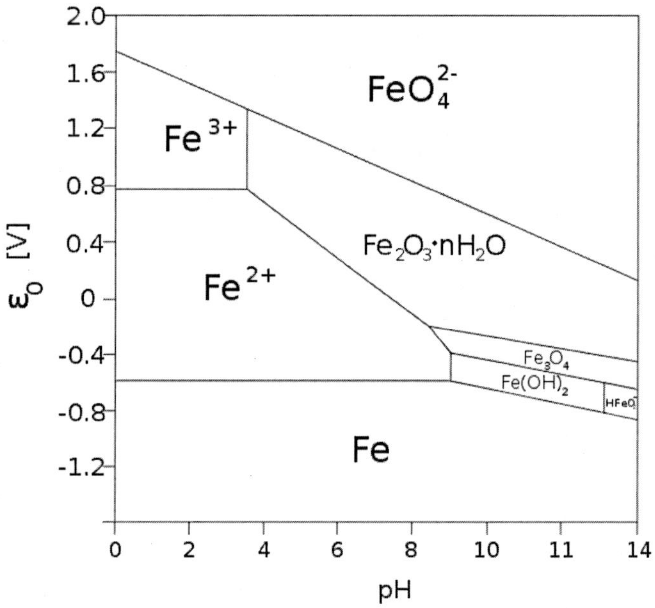

Figure 2 : Pourbaix Diagram of Iron.

The conditions necessary for passivation are recorded in Pourbaix diagrams. Some corrosion inhibitors help the formation of a passivation layer on the surface of the metals to which they are applied. Some compounds, dissolving in solutions (chromates, molybdates) form non-reactive and low solubility films on metal surfaces.

Specific Materials

Silicon

In the area of microelectronics, the formation of a strongly adhering passivating oxide is important to the performance of silicon.

In the area of photovoltaics, a passivating surface layer such as silicon nitride, silicon dioxide or titanium dioxide can reduce surface recombination-a significant loss mechanism in solar cells.

Aluminium

Pure aluminium naturally forms a thin surface layer of aluminium oxide on contact with oxygen in the atmosphere through a process called oxidation, which creates a physical barrier to corrosion or further oxidation in most environments. Aluminium alloys, however, offer little protection against corrosion. There are three main ways to passivate these alloys: *alclading*, *chromate conversion coating* and *anodizing*. Alclading is the process of metallurgically bonding a thin layer of pure aluminium to the aluminium alloy. Chromate conversion coating is a common way of passivating not only aluminum, but also zinc, cadmium, copper, silver, magnesium, and tin alloys. Anodizing forms a thick oxide coating. This finish is more robust than the other processes and also provides good electrical insulation, which the other two processes do not.

For example, prior to storing hydrogen peroxide in an aluminium container, the container can be passivated by rinsing it with a dilute solution of nitric acid and peroxide alternating with deionized water. The nitric acid and peroxide oxidizes and dissolves any impurities on the inner surface of the container, and the deionized water rinses away the acid and oxidized impurities.

Ferrous Materials

Ferrous materials, including steel, may be somewhat protected by promoting oxidation ("rust") and then converting the oxidation to a metalophosphate by using phosphoric acid and further protected by surface coating. As the uncoated surface is water-soluble, a preferred method is to form manganese or zinc compounds by a process commonly known as Parkerizing or phosphate conversion. Older, less-effective but chemically-similar electrochemical conversion coatings included black oxiding, historically known as bluing or browning. Ordinary steel forms a passivating layer in alkali environments, as rebar does in concrete.

A typical passivation process of cleaning stainless steel tanks involves cleaning with sodium hydroxide and citric acid followed by nitric acid (up to 20% at 120 °F) and a complete water rinse. This process will restore the film, remove metal particles, dirt, and welding-generated compounds (e.g. oxides).

Nickel

Nickel can be used for handling elemental fluorine, owing to the formation of a passivation layer of nickel fluoride.

Pitting Corrosion

Pitting corrosion, or pitting, is a form of extremely localized corrosion that leads to the creation of small holes in the metal. The driving power for pitting corrosion is the depassivation of a small area, which becomes anodic while an unknown but potentially vast area becomes cathodic, leading to very localized galvanic corrosion. The corrosion penetrates the mass of the metal, with limited diffusion of ions. The mechanism of pitting corrosion is probably the same as crevice corrosion.

Mechanism

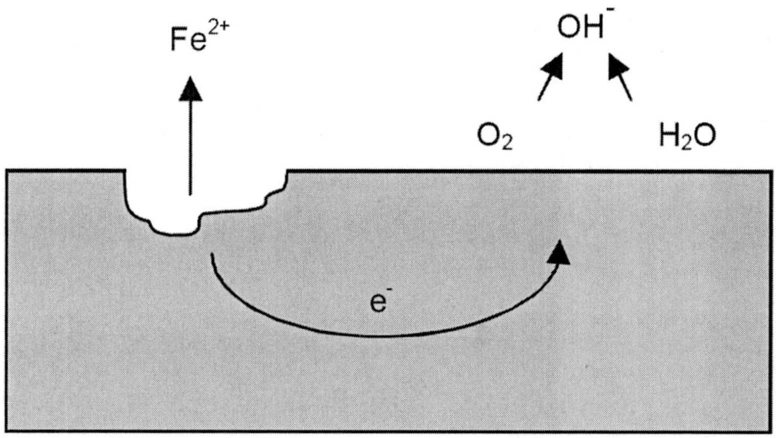

Figure 3 : Diagram Showing a Mechanism of Localized Corrosion Developing on Metal in a solution Containing Oxygen.

It is supposed by some that gravitation causes downward-oriented concentration gradient of the dissolved ions in the hole caused by the corrosion, as the concentrated solution is denser. This however is unlikely. The more conventional explanation is that the acidity inside the pit is maintained by the spatial separation of the cathodic and anodic half-reactions, which creates a potential gradient and electromigration of aggressive anions into the pit.

This kind of corrosion is extremely insidious, as it causes little loss of material with small effect on its surface, while it damages the deep structures of the metal. The pits on the surface are often obscured by corrosion products.

Pitting can be initiated by a small surface defect, being a scratch or a local change in composition, or a damage to protective coating. Polished surfaces display higher resistance to pitting.

Susceptible Alloys

Alloys most susceptible to pitting corrosion are usually the ones where corrosion resistance is caused by a passivation layer: stainless steels, nickel alloys, aluminum alloys. Metals that are susceptible to uniform corrosion in turn do not tend to suffer from pitting. Thus, a regular carbon steel will corrode uniformly in sea water, while stainless steel will pit. Additions of about 2% of molybdenum increases pitting resistance of stainless steels.

Environment

The presence of chlorides, e.g. in sea water, significantly aggravates the conditions for formation and growth of the pits through an autocatalytic process. The pits become loaded with positive metal ions through anodic dissociation.

The Cl^- ions become concentrated in the pits for charge neutrality and encourage the reaction of positive metal ions with water to form a hydroxide corrosion product and H^+ ions. Now, the pits are weakly acidic, which accelerates the process.

Besides chlorides, other anions implicated in pitting include thiosulfates ($S_2O_3^{2-}$), fluorides and iodides. Stagnant water conditions favour pitting. Thiosulfates are particularly aggressive species and are formed by partial oxidation of pyrite, or partial reduction of sulfate. Thiosulfates are a concern for corrosion in many industries: sulfide ores processing, oil wells and pipelines transporting soured oils, Kraft paper production plants, photographic industry, methionine and lysine factories.

Corrosion inhibitors, when present in sufficient amount, will provide protection against pitting. However, too low level of them can aggravate pitting by forming local anodes.

Examples

A single pit in a critical point can cause a great deal of damage. One example is the explosion in Guadalajara, Mexico on April 22, 1992, when gasoline fumes accumulated in sewers destroyed kilometres of streets. The vapours originated from a leak of gasoline through a single hole formed by corrosion between a steel gasoline pipe and a zinc-plated water pipe. Firearms can also suffer from pitting, most notably in the bore of the barrel when corrosive ammunition is used and the barrel is not cleaned soon afterward. Deformities in the bore caused by pitting can greatly reduce the firearms accuracy. To prevent pitting in firearm bores, most modern firearms have a bore lined with chromium.

Pitting corrosion can also help initiate stress corrosion cracking, as happened when a single eyebar on the Silver Bridge, West Virginia and killed 46 people on the bridge in December, 1967.

Intergranular Corrosion

Intergranular corrosion (IGC), also known as intergranular attack (IGA), is a form of corrosion where the boundaries of crystallites of the material are more susceptible to corrosion than their insides. (*Cf.* transgranular corrosion.)

This situation can happen in otherwise corrosion-resistant alloys, when the grain boundaries are depleted, known as *grain boundary depletion*, of the corrosion-inhibiting compound by some mechanism. In nickel alloys and austenitic stainless steels, where chromium is added for corrosion resistance, the mechanism involved is formation of chromium carbide at the grain boundaries, forming chromium-depleted zones (this process is called sensitization). Around 12% chromium is minimally required to ensure passivation, mechanism by which a thin invisible layer forms at the surface of stainless steels. This layer protects the metal from corrosive environments and it is, thus, stainless. Selective leaching often involve grain boundary depletion mechanisms.

These zones also act as local galvanic couples, causing local galvanic corrosion. This condition happens when the material is heated to temperature around 700 °C for too long time, and often happens during welding or an improper heat treatment. When zones of such material form due to welding, the resulting corrosion is termed weld decay. Stainless steels can be stabilized against this behaviour by addition of titanium, niobium, or tantalum, which form titanium carbide, niobium carbide and tantalum carbide preferentially to chromium carbide, by lowering the content of carbon in the steel and in case of welding also in the filler metal under 0.02%, or by heating the entire part above 1000 °C and quenching it in water, leading to dissolution of the chromium carbide in the grains and then preventing its precipitation.

Another possibility is to keep the welded parts thin enough so that, upon cooling, the metal dissipates heat too quickly for chromium carbide to precipitate.

Other related kind of intergranular corrosion is termed knifeline attack (KLA). Knifeline attack impacts steels stabilized by niobium, such as 347 stainless steel. Titanium, niobium, and their carbides dissolve in steel at very high temperatures. At some cooling regimes, niobium carbide does not precipitate, and the steel then behaves like unstabilized steel, forming chromium carbide instead. This affects only a thin zone several millimeters wide in the very vicinity of the weld, making it difficult to spot and increasing the corrosion speed. Structures made of such steels have to be heated in a whole to about 1950 °F, when the chromium carbide dissolves and niobium carbide forms. The cooling rate after this treatment is not important, as the carbon that would otherwise pose risk of formation of chromium carbide is already sequestered as niobium carbide.

Aluminium-based alloys may be sensitive to intergranular corrosion if there are layers of materials acting as anodes between the aluminium-rich crystals. High strength aluminium alloys, especially when extruded or otherwise subjected to high degree of working, can undergo exfoliation corrosion, where the corrosion products build up between the flat, elongated grains and separate them, resulting in lifting or leafing effect and often propagating from edges of the material through its entire structure. Intergranular corrosion is a concern especially for alloys with high content of copper.

Other kinds of alloys can undergo exfoliation as well; the sensitivity of cupronickel increases together with its nickel content. A broader term for this class of corrosion is lamellar corrosion. Alloys of iron are susceptible to lamellar corrosion, as the volume of iron oxides is about seven times higher than the volume of original metal, leading to formation of internal tensile stresses tearing the material apart. Similar effect leads to formation of lamellae in stainless steels, due to the difference of thermal expansion of the oxides and the metal.

Copper-based alloys become sensitive when depletion of copper content in the grain boundaries occurs.

Anisotropic alloys, where extrusion or heavy working leads to formation of long, flat grains, are especially prone to intergranular corrosion.

Intergranular corrosion induced by environmental stresses is termed stress corrosion cracking. Intergranular corrosion can be detected by ultrasonic and eddy current methods.

Sensitization Effect

Certain alloys when exposed to a temperature characterized as a sensitizing temperature become particularly susceptible to intergranular corrosion. In a corrosive atmosphere, the grain interfaces of these sensitized alloys become very reactive and intergranular corrosion results. This is characterized by a localized attack at an adjacent to grain boundaries with relatively little corrosion of the grains themselves. The alloy disintegrates (grains fall out) and/or loses its strength.

Intergranular corrosion is generally considered to be caused by the segregation of impurities at the grain boundaries or by enrichment or depletion of one of the alloying elements in the grain boundary areas. Thus in certain aluminium alloys, small amounts of iron have been shown to segregate in the grain boundaries and cause intergranular corrosion.

Also, it has been shown that the zinc content of a brass is higher at the grain boundaries and subject to such corrosion. High-strength aluminium alloys such as the Duralumin-type alloys (Al-Cu) which depend upon precipitated phases for strengthening are susceptible to intergranular corrosion following sensitization at temperatures of about 120°C. Nickel-rich alloys such as Inconel 600 and Incoloy 800 show similar susceptibility.

Die-cast zinc alloys containing aluminum exhibit intergranular corrosion by steam in a marine atmosphere. Cr-Mn and Cr-Mn-Ni steels are also susceptible to intergranular corrosion following sensitization in the temperature range of 400°-850°C. In the case of the austenitic stainless steels, when these steels are sensitized by being heated in the temperature range of about 500° to 800°C, depletion of chromium in the grain boundary region occurs, resulting in susceptibility to intergranular corrosion. Such sensitization of austenitic stainless steels can readily occur because of temperature service requirements, as in steam generators, or as a result of subsequent welding of the formed structure.

The following photos show the typical microstructure of a normalized (unsensitized) type 304 stainless steel and a heavily sensitized steel. The dark lines in the sensitized microstructure are networks of chromium carbides precipitated along the grain boundaries.

Several methods have been used to control or minimize the intergranular corrosion of susceptible alloys, particularly of the austenitic stainless steels. Thus a high-temperature solution heat treatment, commonly termed solution-annealing, quench-annealing or solution-quenching, has been used. The alloy is heated to a temperature of about 1,060° to 1,120°C and then water quenched. This method is generally unsuitable for treating large assemblies, and also ineffective where welding is subsequently used for making repairs or for attaching other structures. Another control technique for preventing intergranular corrosion involves incorporating strong carbide formers or stabilizing elements such as niobium or titanium in the stainless steels.

Such elements have a much greater affinity for carbon than does chromium; carbide formation with these elements reduces the carbon available in the alloy for formation of chromium carbides. Such a stabilized titanium-bearing austenitic chromium-nickel-copper stainless steel is shown in U.S. Pat. No. 3,562,781. Or the stainless steel may initially be reduced in carbon content below 0.03 percent so that insufficient carbon is provided for carbide formation. These techniques are expensive and only partially effective since sensitization may occur with time. The low-carbon steels also frequently exhibit lower strengths at high temperatures.

Crevice Corrosion

Crevice corrosion is a corrosion occurring in spaces to which the access of the working fluid from the environment is limited. These spaces are generally called crevices. Examples of crevices are gaps and contact areas between parts, under gaskets or seals, inside cracks and seams, spaces filled with deposits and under sludge piles.

This photo shows that corrosion occurred in the crevice between the tube and tube sheet (both made of type 316 stainless steel) in a sea water desalination plant.

Mechanism

The corrosion resistance of a stainless steel is dependent on the presence of a protective oxide layer on its surface, but it is possible under certain conditions for this oxide layer to break down, for example in reducing acids, or in some types of combustion where the atmosphere is reducing. Areas where the oxide layer can break down can also sometimes be the result of the way components are designed, for example under gaskets, in sharp re-entrant corners or associated with incomplete weld penetration or overlapping surfaces. These can all form crevices which can promote corrosion. To function as a corrosion site, a crevice has to be of sufficient width to permit entry of the corrodent, but sufficiently narrow to ensure that the corrodent remains stagnant. Accordingly crevice corrosion usually occurs in gaps a few micrometres wide, and is not found in grooves or slots in which circulation of the corrodent is possible. This problem can often be overcome by paying attention to the design of the component, in particular to avoiding formation of crevices or at least keeping them as open as possible. Crevice corrosion is a very similar mechanism to pitting corrosion; alloys resistant to one are generally resistant to both. Crevice corrosion can be viewed as a more severe form of pitting corrosion as it will occur at significantly lower temperatures than does pitting.

Crevices can develop a local chemistry which is very different from that of the bulk fluid. For example, in boilers, concentration of non-volatile impurities may occur in crevices near heat-transfer surfaces because of the continuous water vaporization. "Concentration factors" of many millions are not uncommon for common water impurities like sodium, sulfate or chloride. The concentration process is often referred to as "hideout" (HO), whereas the opposite process, whereby the concentrations tend to even out (e.g., during shutdown) is called "hideout return" (HOR).

Two factors are important in the initiation of active crevice corrosion: the chemical composition of the electrolyte in the crevice and the potential drop into the crevice. Researchers had previously claimed that either one or the other of the two factors was responsible for initiating crevice corrosion, but recently it has been shown that it is a combination of the two that causes active crevice corrosion. Both the potential drop and the change in composition of the crevice electrolyte are caused by deoxygenation of the crevice and a separation of electroactive areas, with net anodic reactions occurring within the crevice and net cathodic reactions occurring exterior to the crevice (on the bold surface). The ratio of the surface areas between the cathodic and anodic region is significant.

Some of the phenomena occurring within the crevice may be somewhat reminiscent of galvanic corrosion:

- galvanic corrosion
- two connected metals + single environment
- crevice corrosion
- one metal part + two connected environments.

The mechanism of crevice corrosion can be (but is not always) similar to that of pitting corrosion. However, there are sufficient differences to warrant a separate treatment. For example, in crevice corrosion, one has to consider the geometry of the crevice and the nature of the concentration process leading to the development of the differential local chemistry. The extreme and often unexpected local chemistry conditions inside the crevice need to be considered. Galvanic effects can play a role in crevice degradation.

Mode of attack

Depending on the environment developed in the crevice and the nature of the metal, the crevice corrosion can take a form of:

- pitting (i.e., formation of pits),
- filiform corrosion (this type of crevice corrosion that may occur on an aluminium surface underneath an organic coating),
- intergrannular attack, or
- stress corrosion cracking.

Stress Corrosion Cracking

A common form of crevice failure occurs due to stress corrosion cracking, where a crack or cracks develop from the base of the crevice where the stress concentration is greatest. This was the root cause of the fall of the Silver Bridge in 1967 in West Virginia, where a single critical crack only about 3 mm long suddenly grew and fractured a tie bar joint. The rest of the bridge fell in less than a minute. The eyebars in the Silver Bridge were not redundant, as links were composed of only two bars each, of high strength steel (more than twice as strong as common mild steel), rather than a thick stack of thinner bars of modest material strength "combed" together as is usual for redundancy. With only two bars, the failure of one could impose excessive loading on the second, causing total failure—unlikely if more bars are used. While a low-redundancy chain can be engineered to the design requirements, the safety is completely dependent upon correct, high quality manufacturing and assembly.

Significance

The susceptibility to crevice corrosion varies widely from one material-environment system to another. In general, crevice corrosion is of greatest concern for materials which are normally passive metals, like stainless steel or aluminum. Crevice corrosion tends to be of greatest significance to components built of highly corrosion-resistant superalloys and operating with the purest-available water chemistry. For example, steam generators in nuclear power plants degrade largely by crevice corrosion. Crevice corrosion is extremely dangerous because it is localized and can lead to component failure while the overall material loss is minimal. The initiation and progress of crevice corrosion can be difficult to detect.

Microbial Corrosion

Microbial corrosion, also called bacterial corrosion, bio-corrosion, microbiologically-influenced corrosion, or microbially-induced corrosion (MIC), is corrosion caused or promoted by microorganisms, usually chemoautotrophs. It can apply to both metals and non-metallic materials.

Bacteria

Some sulfate-reducing bacteria produce hydrogen sulfide, which can cause sulfide stress cracking. *Acidithiobacillus* bacteria produce sulfuric acid; *Acidothiobacillus thiooxidans* frequently damages sewer pipes. *Ferrobacillus ferrooxidans* directly oxidizes iron to iron oxides and iron hydroxides; the rusticles forming on RMS *Titanic* wreck are caused by bacterial activity. Other bacteria produce various acids, both organic and mineral, or ammonia.

In presence of oxygen, aerobic bacteria like *Acidithiobacillus thiooxidans*, *Thiobacillus thioparus*, and *Thiobacillus concretivorus*, all three widely present in the environment, are the common corrosion-causing factors resulting in biogenic sulfide corrosion.

Without presence of oxygen, anaerobic bacteria, especially *Desulfovibrio* and *Desulfotomaculum*, are common. *Desulfovibrio salixigens* requires at least 2.5% concentration of sodium chloride, but *D. vulgaris* and *D. desulfuricans* can grow in both fresh and salt water. *D. africanus* is another common corrosion-causing microorganism.

The *Desulfotomaculum* genus comprises sulfate-reducing spore-forming bacteria; *Dtm. orientis* and *Dtm. nigrificans* are involved in corrosion processes. Sulfate-reducers require reducing environment; the electrode potential of at least-100 mV is required for them to thrive. However, even a small amount of produced hydrogen sulfide can achieve this shift, so the growth, once started, tends to accelerate.

Layers of anaerobic bacteria can exist in the inner parts of the corrosion deposits, while the outer parts are inhabited by aerobic bacteria.

Some bacteria are able to utilize hydrogen formed during cathodic corrosion processes.

Bacterial colonies and deposits can form concentration cells, causing and enhancing galvanic corrosion.

Bacterial corrosion may appear in form of pitting corrosion, for example in pipelines of the oil and gas industry. Anaerobic corrosion is evident as layers of metal sulfides and hydrogen sulfide smell. On cast iron, a graphitic corrosion selective leaching may be the result, with iron being consumed by the bacteria, leaving graphite matrix with low mechanical strength in place.

Various corrosion inhibitors can be used to combat microbial corrosion. Formulae based on benzalkonium chloride are common in oilfield industry.

Microbial corrosion can also apply to plastics, concrete, and many other materials. Two examples are Nylon-eating bacteria and Plastic-eating bacteria.

Aviation Fuel

Hydrocarbon utilizing microorganisms, mostly *Cladosporium resinae* and *Pseudomonas aeruginosa*, colloquially known as "HUM bugs", are commonly present in jet fuel. They live in the water-fuel interface of the water droplets, form dark black/brown/green, gel-like mats, and cause microbial corrosion to plastic and rubber parts of the aircraft fuel system by consuming them, and to the metal parts by the means of their acidic metabolic products. They are also incorrectly called algae due to their appearance. FSII, which is added to the fuel, acts as a growth retardant for them. There are about 250 kinds of bacteria that can live in jet fuel, but fewer than a dozen are meaningfully harmful.

Recent Events

In response to increased awareness of the nature and danger of microbial corrosion, a two-day international symposium was held in Perth, Western Australia in February 2007. The symposium, originally proposed by Dr. Reza Javaherdashti, was sponsored by EXTRIN Consultants and Curtin University of Technology, as well as other local industries. It attracted speakers and attendees from as far as Argentina, Brazil, New Zealand, the UK and the US in addition to Australian representatives. The symposium primarily focussed on the identification of Microbial Corrosion in marine, mining and industrial environments and the best course of action to remove and prevent further attacks.

High Temperature Corrosion

High temperature corrosion is chemical deterioration of a material (typically a metal) under very high temperature conditions. This non-galvanic form of corrosion can occur when a metal is subject to a high temperature atmosphere containing oxygen, sulfur or other compounds capable of oxidising (or assisting the oxidation of) the material concerned. For example, materials used in aerospace, power generation and even in car engines have to resist sustained periods at high temperature in which they may be exposed to an atmosphere containing potentially highly corrosive products of combustion.

The products of high temperature corrosion can potentially be turned to the advantage of the engineer. The formation of oxides on stainless steels, for example, can provide a protective layer preventing further atmospheric attack, allowing for a material to be used for sustained periods at both room and high temperature in hostile conditions. Such high temperature corrosion products in the form of compacted oxide layer glazes have also been shown to prevent or reduce wear during high temperature sliding contact of metallic (or metallic and ceramic) surfaces.

Chapter 7

Galvanization

Galvanization (or galvanisation) is the process of applying a protective zinc coating to steel or iron, in order to prevent rusting. The term is derived from the name of Italian scientist Luigi Galvani.

Although galvanization can be done with electrochemical and electrodeposition processes, the most common method in current use is hot-dip galvanization, in which steel parts are submerged in a bath of molten zinc.

Meanings

Metal Protection

In current use, the term refers to the coating of steel or iron with zinc. This is done to prevent rusting of the ferrous item. The value of galvanizing stems from the relative corrosion resistance of zinc, which, under most service conditions, is considerably less than those of iron and steel. The zinc therefore serves as a sacrificial anode, so that it cathodically protects exposed steel. This means that even if the coating is scratched or abraded, the exposed steel will still be protected from corrosion by the remaining zinc-an advantage absent from paint, enamel, powder coating and other methods. Galvanizing is also favoured as a means of protective coating because of its low cost, ease of application and comparatively long maintenance-free service life.

The term galvanizing, while technically referring specifically to the application of zinc coating by the use of a galvanic cell (also known as electroplating), is also generally understood to include hot-dip zinc coating. The practical difference is that hot-dip galvanization produces a thick, durable and matte gray coating-electroplated coatings tend to be thin and brightly reflective. Due to its thinness, the zinc of electroplated coatings is quickly depleted, making them unsuitable for outdoor applications (except in very dry climates). When combined with subsequent painting (which slows zinc consumption), electroplating is durable enough to be used in some premium auto body coatings.

Nonetheless, electroplating is used on its own for many outdoor applications because it is cheaper than hot dip zinc coating and looks good when new. Another reason not to use hot dip zinc coating is that for bolts and nuts size M10 or smaller, the thick hot-dipped coating uses up too much of the threads, which reduces strength (because the dimension of the steel prior to coating must be reduced for the fasteners to fit together).

This means that for cars, bicycles and many other 'light' mechanical products, the alternative to electroplating bolts and nuts is not hot dip zinc coating but making the bolts and nuts from stainless steel (known by the corrosion grades A4 and A2).

Electroplated steel is visually indistinguishable from stainless steel when new. To determine whether a part is electroplated or stainless steel, apply a magnet. The most common stainless steel alloys (including those used for bolts and nuts) are not magnetic or only very slightly attracted to a magnet.

History

Originally, "galvanization" was the administration of electric shocks (in the 19th century also termed *Faradism*, after Michael Faraday). It stemmed from Galvani's induction of twitches in severed frogs' legs, by his accidental generation of electricity. Its claims to health benefits have largely been disproved, except for some limited uses in psychiatry in the form of electroconvulsive therapy (ECT).

This archaic sense is the origin of the meaning of *galvanic* when meaning "affected/affecting, as if by a shock of electricity; startled". and the metaphorical "galvanize into action" referring to suddenly stimulating a complacent person or group to take action. Later the word was used for processes of electrodeposition. This remains a useful and broadly applied technology, but the term "galvanization" has largely come to be associated with zinc coatings, to the exclusion of other metals.

Galvanic paint, a precursor to hot-dip galvanization, was patented by Stanislas Sorel, of Paris, France in December, 1837.

The earliest known example of galvanizing of iron was found on the 30th September 1999 by the Royal Armouries Museum on a 17th century Indian armour in their collection.

Zinc Coatings

Zinc coatings prevent corrosion of the protected metal by forming a physical barrier, and by acting as a sacrificial anode if this barrier is damaged. When exposed to the atmosphere, zinc reacts with oxygen to form zinc oxide, which further reacts with water molecules in the air to form zinc hydroxide. Finally zinc hydroxide reacts with carbon dioxide in the atmosphere to yield a thin, impermeable, tenacious and quite insoluble dull gray layer of zinc carbonate which adheres extremely well to the underlying zinc, so protecting it from further corrosion, in a way similar to the protection afforded to aluminium and stainless steels by their oxide layers.

Hot-dip galvanizing deposits a thick robust layer that may be more than is necessary for the protection of the underlying metal in some applications. This is the case in automobile bodies, where additional rust proofing paint will be applied. Here, a thinner form of galvanizing is applied by electroplating, called “electrogalvanization”. The hot-dip process slightly reduces the strength of the base metal, which is a consideration for the manufacture of wire rope and other highly-stressed products. The protection provided by this process is insufficient for products that will be constantly exposed to corrosive materials such as salt water. For these applications, more expensive stainless steel is preferred. Some nails made today are electro-galvanized.

As noted previously, both mechanisms are often at work in practical applications. For example, the traditional measure of a coating's effectiveness is resistance to a salt spray. Thin coatings cannot remain intact indefinitely when subject to surface abrasion, and the galvanic protection offered by zinc can be sharply contrasted to more noble metals. As an example, a scratched or incomplete coating of chromium actually exacerbates corrosion of the underlying steel, since it is less electrochemically active than the substrate.

The size of crystallites in galvanized coatings is an aesthetic feature, known as spangle. By varying the number of particles added for heterogeneous nucleation and the rate of cooling in a hot-dip process, the spangle can be adjusted from an apparently uniform surface (crystallites too small to see with the naked eye) to grains several centimetres wide. Visible crystallites are rare in other engineering materials. Protective coatings for steel constitute the largest use of zinc and rely upon the galvanic or sacrificial property of zinc relative to steel.

Thermal diffusion galvanizing, a form of Sherardizing, provides a zinc coating on iron or copper based materials partially similar to hot dip galvanizing. The final surface is different than hot-dip Galvanizing; all of its zinc is alloyed. Zinc is applied in a powder form with "accelerator chemicals" (generally sand, but other chemicals are patented).

The parts and the zinc powder are tumbled in a sealed drum while it is heated to slightly below zinc's melting temperature. The drum must be heated evenly, or complications will arise. Due to the chemicals added to the zinc powder, the zinc/iron makes an alloy at a lower temperature than hot dip galvanizing.

This process requires generally fewer preparatory cleanings than other methods. The dull-grey crystal structure formed by the process bonds stronger with paint, powder coating, and rubber overmolding processes than other methods. It is a preferred method for coating small, complex-shaped metals and smoothing in rough surfaces on items formed with powder metal.

Eventual Corrosion

Although galvanizing will inhibit attack of the underlying steel, rusting will be inevitable, especially due to natural acidity of rain. For example, corrugated iron sheet roofing will start to degrade within a few years despite the protective action of the zinc coating. Marine and salty environments also lower the lifetime of galvanized iron because the high electrical conductivity of sea water increases the rate of corrosion. Galvanized car frames exemplify this; they corrode much quicker in cold environments due to road salt. Galvanized steel can last for many years if other means are maintained, such as paint coatings and additional sacrificial anodes.

Reactive Coatings

If the environment is controlled (especially in recirculating systems), corrosion inhibitors can often be added to it. These form an electrically insulating and/or chemically impermeable coating on exposed metal surfaces, to suppress electrochemical reactions. Such methods obviously make the system less sensitive to scratches or defects in the coating, since extra inhibitors can be made available wherever metal becomes exposed. Chemicals that inhibit corrosion include some of the salts in hard water (Roman water systems are famous for their mineral deposits), chromates, phosphates, polyaniline, other conducting polymers and a wide range of specially-designed chemicals that resemble surfactants (i.e. long-chain organic molecules with ionic end groups).

Anodizing

Anodizing, or anodising in British English, is an electrolytic passivation process used to increase the thickness of the natural oxide layer on the surface of metal parts. The process is called "anodizing" because the part to be treated forms the anode electrode of an electrical circuit. Anodizing increases corrosion resistance and wear resistance, and provides better adhesion for paint primers and glues than does bare metal. Anodic films can also be used for a number of cosmetic effects, either with thick porous coatings that can absorb dyes or with thin transparent coatings that add interference effects to reflected light. Anodizing is also used to prevent galling of threaded components and to make dielectric films for electrolytic capacitors.

Anodic films are most commonly applied to protect aluminium alloys, although processes also exist for titanium, zinc, magnesium, niobium, and tantalum. Iron or carbon steel metal exfoliates when oxidized under neutral or alkaline microelectrolytic conditions; i.e., the iron oxide (actually "ferric hydroxide" or hydrated iron oxide, also known as rust) forms by anoxic anodic pits and large cathodic surface, these pits concentrate anions such as sulfate and chloride accelerating the underlying metal to corrosion. Carbon flakes or nodules in iron or steel with high carbon content (high carbon steel, cast iron) may cause an electrolytic potential and interfere with coating or plating. Ferrous metals are commonly anodized electrolytically in nitric acid, or by treatment with red fuming nitric acid, to form hard black ferric oxide. This oxide remains conformal even when plated on wire and the wire is bent.

Anodization changes the microscopic texture of the surface and changes the crystal structure of the metal near the surface. Thick coatings are normally porous, so a sealing process is often needed to achieve corrosion resistance. Anodized aluminium surfaces, for example, are harder than aluminium but have low to moderate wear resistance that can be improved with increasing thickness or by applying suitable sealing substances. Anodic films are generally much stronger and more adherent than most types of paint and metal plating, but also more brittle. This makes them less likely to crack and peel from aging and wear, but more susceptible to cracking from thermal stress.

History

Anodizing was first used on an industrial scale in 1923 to protect Duralumin seaplane parts from corrosion. This early chromic acid-based process was called the Bengough-Stuart process and was documented in British defence specification DEF STAN 03-24/3. It is still used today despite its legacy requirements for a complicated voltage cycle now known to be unnecessary. Variations of this process soon evolved, and the first sulfuric acid anodizing process was patented by Gower and O'Brien in 1927. Sulfuric acid soon became and remains the most common anodizing electrolyte.

Oxalic acid anodizing was first patented in Japan in 1923 and later widely used in Germany, particularly for architectural applications. Anodized aluminium extrusion was a popular architectural material in the 1960s and 1970s, but has since been displaced by cheaper plastics and powdercoating. The phosphoric acid processes are the most recent major development, so far only used as pretreatments for adhesives or organic paints. A wide variety of proprietary and increasingly complex variations of all these anodizing processes continue to be developed by industry, so the growing trend in military and industrial standards is to classify by coating properties rather than by process chemistry.

Anodized Aluminium

Aluminium alloys are anodized to increase corrosion resistance, to increase surface hardness, and to allow dyeing (colouring), improved lubrication, or improved adhesion. The anodic layer is non-conductive.

When exposed to air at room temperature, or any other gas containing oxygen, pure aluminium self-passivates by forming a surface layer of amorphous aluminium oxide 2 to 3 nm thick, which provides very effective protection against corrosion. Aluminium alloys typically form a thicker oxide layer, 5-15 nm thick, but tend to be more susceptible to corrosion. Aluminium alloy parts are anodized to greatly increase the thickness of this layer for corrosion resistance. The corrosion resistance of aluminium alloys is significantly decreased by certain alloying elements or impurities: copper, iron, and silicon, so 2000, 4000, and 6000-series alloys tend to be most susceptible.

Although anodizing produces a very regular and uniform coating, microscopic fissures in the coating can lead to corrosion. Further, the coating is susceptible to chemical dissolution in the presence of high and low pH chemistry, which results in stripping the coating and corrosion of the substrate. To combat this, various techniques have been developed either to reduce the number of fissures or to insert more chemically stable compounds into the oxide, or both. For instance, sulfuric anodized articles are normally sealed, either through hydrothermal sealing or precipitating sealing, to reduce porosity and interstitial pathways that allow for corrosive ion exchange between the surface and the substrate. Precipitating seals enhance chemical stability but are less effective in eliminating ion exchange pathways. Most recently, new techniques to partially convert the amorphous oxide coating into more stable micro-crystalline compounds have been developed that have shown significant improvement based on shorter bond lengths.

Some aluminium aircraft parts, architectural materials, and consumer products are anodized. Anodized aluminium can be found on mp3 players, flashlights, cookware, cameras, sporting goods, window frames, roofs, in electrolytic capacitors, and on many other products both for corrosion resistance and the ability to retain dye. Although anodizing only has moderate wear resistance, the deeper pores can better retain a lubricating film than a smooth surface would.

Anodized coatings have a much lower thermal conductivity and coefficient of linear expansion than aluminium. As a result, the coating will crack from thermal stress if exposed to temperatures above 80 °C. The coating can crack, but it will not peel. The melting point of aluminium oxide is 2050 °C, much higher than pure aluminium's 658 °C. (This can make welding more difficult.) In typical commercial aluminium anodization processes, the aluminium oxide is grown down into the surface and out from the surface by equal amounts. So anodizing will increase the part dimensions on each surface by half of the oxide thickness. For example a coating that is (2 μm) thick, will increase the part dimensions by (1 μm) per surface. If the part is anodized on all sides, then all linear dimensions will increase by the oxide thickness. Anodized aluminium surfaces are harder than aluminium but have low to moderate wear resistance, although this can be improved with thickness and sealing.

Process

Preceding the anodization process, wrought alloys are cleaned in either a hot soak cleaner or in a solvent bath and may be etched in sodium hydroxide (normally with added sodium gluconate), ammonium bifluoride or brightened in a mix of acids. Cast alloys are normally best just cleaned due to the presence of intermetallic substances unless they are a high purity alloy such as LM0.

The anodized aluminium layer is grown by passing a direct current through an electrolytic solution, with the aluminium object serving as the anode (the positive electrode).

The current releases hydrogen at the cathode (the negative electrode) and oxygen at the surface of the aluminium anode, creating a build-up of aluminium oxide. Alternating current and pulsed current is also possible but rarely used.

The voltage required by various solutions may range from 1 to 300 V DC, although most fall in the range of 15 to 21 V. Higher voltages are typically required for thicker coatings formed in sulfuric and organic acid. The anodizing current varies with the area of aluminium being anodized, and typically ranges from 30 to 300 amperes/meter^2 (2.8 to 28 ampere/ft^2).

Aluminium anodizing is usually performed in an acid solution which slowly dissolves the aluminium oxide. The acid action is balanced with the oxidation rate to form a coating with nanopores, 10-150 nm in diameter. These pores are what allows the electrolyte solution and current to reach the aluminium substrate and continue growing the coating to greater thickness beyond what is produced by autopassivation. However, these same pores will later permit air or water to reach the substrate and initiate corrosion if not sealed. They are often filled with coloured dyes and/or corrosion inhibitors before sealing. Because the dye is only superficial, the underlying oxide may continue to provide corrosion protection even if minor wear and scratches may break through the dyed layer.

Conditions such as electrolyte concentration, acidity, solution temperature, and current must be controlled to allow the formation of a consistent oxide layer. Harder, thicker films tend to be produced by more dilute solutions at lower temperatures with higher voltages and currents. The film thickness can range from under 0.5 micrometers for bright decorative work up to 150 micrometers for architectural applications.

Other Widely used Specifications

The most widely used anodizing specification is a U.S. military spec, MIL-A-8625, which defines three types of aluminium anodization. Type I is chromic acid anodization, Type II is sulfuric acid anodization, and Type III is sulfuric acid hardcoat anodization. Other anodizing specifications include more MIL-SPECs (e.g., MIL-A-63576), aerospace industry specs by organizations such as SAE, ASTM, and ISO (e.g., AMS 2469, AMS 2470, AMS 2471, AMS 2472, AMS 2482, ASTM B580, ASTM D3933, ISO 10074, and BS 5599), and corporation-specific specs (such as those of Boeing, Lockheed Martin, and other large contractors). AMS 2468 is obsolete. None of these specifications define a detailed process or chemistry, but rather a set of tests and quality assurance measures which the anodized product must meet. BS 1615 provides guidance in the selection of alloys for anodizing. For British defence work, a detailed chromic and sulfuric anodizing processes are described by STAN 03-24/3 and STAN 03-25/3 respectively.

Chromic acid Anodizing (Type I)

The oldest anodizing process uses chromic acid. It is widely known as the Bengough-Stuart process. In North America it is known as Type I because it is so designated by the MIL-A-8625 standard, but it is also covered by AMS 2470 and MIL-A-8625 Type IB. In the UK it is normally specified as Def Stan 03/24 and used in areas that are prone to come into contact with propellants etc. There are also Boeing and Airbus standards. Chromic acid produces thinner, 0.5 μm to 18 μm (0.00002 to 0.0007) more opaque films that are softer, ductile, and to a degree self-healing. They are harder to dye and may be applied as a pretreatment before painting. The method of film formation is different from using sulfuric acid in that the voltage is ramped up through the process cycle.

Sulfuric Acid Anodizing (Type II & III)

Sulfuric acid is the most widely used solution to produce anodized coating. Coatings of moderate thickness 1.8 μm to 25 μm (0.00007 to 0.001) are known as Type II in North America, as named by MIL-A-8625, while coatings thicker than 25 μm (0.001) are known as Type III, hardcoat, hard anodizing, or engineered anodizing. Very thin coatings similar to those produced by chromic anodizing are known as Type IIB. Thick coatings require more process control, and are produced in a refrigerated tank near the freezing point of water with higher voltages than the thinner coatings. Hard anodizing can be made between 13 and 150 μm (0.0005 to 0.006) thick. Anodizing thickness increases wear resistance, corrosion resistance, ability to retain lubricants and PTFE coatings, and electrical and thermal insulation. Standards for thin (Soft/Standard) sulfuric anodizing are given by MIL-A-8625 Types II and IIB, AMS 2471 (undyed), and AMS 2472 (dyed), BS EN ISO 12373/1 (decorative), BS EN 3987 (Architectural). Standards for thick sulfuric anodizing are given by MIL-A-8625 Type III, AMS 2469, BS 5599, BS EN 2536 and the obsolete AMS 2468 and DEF STAN 03-26/1.

Organic Acid Anodizing

Anodizing can produce yellowish integral colours without dyes if it is carried out in weak acids with high voltages, high current densities, and strong refrigeration. Shades of colour are restricted to a range which includes pale yellow, gold, deep bronze, brown, grey, and black. Some advanced variations can produce a white coating with 80% reflectivity. The shade of colour produced is sensitive to variations in the metallurgy of the underlying alloy and cannot be reproduced consistently.

Anodization in some organic acids, for example Malic Acid, can enter a 'runaway' situation, in which the current drives the acid to attack the aluminum far more aggressively than normal, resulting in huge pits and scarring. Also, if the current or voltage are driven too high, 'burning' can set in; in this case the supplies act as if nearly shorted and large, uneven and amorphous black regions develop.

Integral colour anodizing is generally done with organic acids, but the same effect has been produced in laboratory with very dilute sulfuric acid. Integral colour anodizing was originally performed with oxalic acid, but sulfonated aromatic compounds containing oxygen, particularly sulfosalicylic acid, have been more common since the 1960s. Thicknesses up to 50μm can be achieved. Organic acid anodizing is called Type IC by MIL-A-8625.

Phosphoric Acid Anodizing

Anodizing can be carried out in phosphoric acid, usually as a surface preparation for adhesives. This is described in standard ASTM D3933.

Borate and Tartrate Baths

Anodizing can also be performed in borate or tartrate baths in which aluminium oxide is insoluble. In these processes, the coating growth stops when the part is fully covered, and the thickness is linearly related to the voltage applied. These coatings are free of pores, relative to the sulfuric and chromic acid processes. This type of coating is widely used to make electrolytic capacitors, because the thin aluminium films (typically less than 0.5 μm) would risk being pierced by acidic processes.

Plasma Electrolytic Oxidation

Plasma electrolytic oxidation is a similar process, but where higher voltages are applied. This causes sparks to occur, and results in more crystalline/ceramic type coatings.

Other Metals

Anodized Titanium

Anodized titanium is used in a recent generation of dental implants. An anodized oxide layer has a thickness in the range of 30 nanometres (1.2×10^{-6} in) to several micrometers. Standards for titanium anodizing are given by AMS 2487 and AMS 2488.

Anodizing titanium generates an array of different colours without dyes, for which it is sometimes used in art, costume jewellery, body piercing jewellery and wedding rings. The colour formed is dependent on the thickness of the oxide (which is determined by the anodizing voltage); it is caused by the interference of light reflecting off the oxide surface with light travelling through it and reflecting off the underlying metal surface. Titanium nitride coatings can also be formed, which have a brown or golden colour and have the same wear and corrosion benefits as anodization.

Anodized Magnesium

Magnesium is anodized primarily as a primer for paint. A thin (5μm) film is sufficient for this. Thicker coatings of 25μm and up can provide mild corrosion resistance when sealed with oil, wax, or sodium silicate. Standards for magnesium anodizing are given in AMS 2466, AMS 2478, AMS 2479, and ASTM B893.

Anodized Zinc

Zinc is rarely anodized, but a process was developed by the International Lead Zinc Research Organization and covered by MIL-A-81801. A solution of ammonium phosphate, chromate and fluoride with voltages of up to 200V can produce olive green coatings up to 80μm thick. The coatings are hard and corrosion resistant.

Anodized Niobium

Niobium anodizes in a similar fashion to titanium with a range of attractive colours being formed by interference at different film thicknesses. Again the film thickness is dependent on the anodising voltage. Uses include jewellery and commemorative coins.

Anodized Tantalum

Tantalum anodizes in a similar fashion to titanium and niobium with a range of attractive colours being formed by interference at different film thicknesses. Again the film thickness is dependent on the anodizing voltage and typically ranges from 18-23 Angstroms per volt depending on electrolyte and temperature. Uses include Tantalum capacitors

Dyeing

The most common anodizing processes, for example sulfuric acid on aluminium, produce a porous surface which can accept dyes easily. The number of dye colours is almost endless; however, the colours produced tend to vary according to the base alloy.

Though some may prefer lighter colours, in practice they may be difficult to produce on certain alloys such as high-silicon casting grades and 2000-series aluminium-copper alloys.

Another concern is the "lightfastness" of organic dyestuffs—some colours (reds and blues) are particularly prone to fading. Black dyes and gold produced by inorganic means (ferric ammonium oxalate) are more lightfast. Dyed anodizing is usually sealed to reduce or eliminate dye bleed out.

Alternatively, metal (usually tin) can be electrolytically deposited in the pores of the anodic coating to provide colours that are more lightfast. Metal dye colours range from pale champagne to black. Bronze shades are commonly used for architectural use.

Alternatively the colour may be produced integral to the film. This is done during the anodizing process using organic acids mixed with the sulfuric electrolyte and a pulsed current.

Splash effects are created by dying the unsealed porous surface in lighter colours and then splashing darker colour dyes onto the surface. Aqueous and solvent based dye mixtures may also be alternately applied since the coloured dyes will resist each other and leave spotted effects.

Printing

Photo quality images and graphics in vivid colour may be printed into the unsealed porous oxide layer using colour dyes via silkscreen, sublimation transfer or digital printer. Line art quality graphics can be achieved by use of a printer. Colour graphics may also be directly applied by hand using an airbrush, sponge or paintbrush. Printed anodizing is sealed to prevent or reduce dye bleed out. Uses include baseball bats, signs, furniture, surgical trays, motorcycle components, and architectural moulding.

Sealing

Acidic anodizing solutions produce pores in the anodized coating. These pores can absorb dyes and retain lubricants, but are also an avenue for corrosion. When lubrication properties are not critical, they are usually sealed after dyeing to increase corrosion resistance and dye retention.

Long immersion in boiling-hot deionized water or steam is the simplest sealing process, although it is not completely effective and reduces abrasion resistance by 20%. The oxide is converted into its hydrated form, and the resulting swelling reduces the porosity of the surface. Cold sealing, where the pores are closed by impregnation of a sealant in a room-temperature bath, is more popular due to energy savings. Coatings sealed in this method are not suitable for adhesive bonding. Teflon, nickel acetate, cobalt acetate, and hot sodium or potassium dichromate seals are commonly used. MIL-A-8625 requires sealing for thin coatings (Types I and II) and allows it as an option for thick ones (Type III).

Cleaning

Anodized aluminium surfaces are susceptible to Panel Edge Staining, a unique type of surface staining that can affect the structural integrity of the metal.

Environmental Impact

Anodizing is one of the more environmentally friendly metal finishing processes. With the exception of organic (aka integral colour) anodizing, the by-products do not contain heavy metals, halogens or volatiles. The most common anodizing effluents, aluminium hydroxide and aluminium sulfate, are recycled for the manufacturing of alum, baking powder, cosmetics, newsprint and fertilizer or used by industrial wastewater treatment systems.

Mechanical Considerations

Anodizing will raise the surface, since the oxide created occupies more space than the base metal converted. This will generally not be of consequence except in the case of small holes threaded to accept screws. Anodizing may cause the screws to bind, thus the threaded holes may need to be chased with a tap to restore the original dimensions.

Alternatively, special oversize taps may be used to precompensate for this growth. In the case of unthreaded holes that accept fixed diameter pins or rods a slightly oversized hole to allow for the dimension change may be appropriate.

Biofilm Coatings

A new form of protection has been developed by applying certain species of bacterial films to the surface of metals in highly corrosive environments. This process increases the corrosion resistance substantially. Alternatively, antimicrobial-producing biofilms can be used to inhibit mild steel corrosion from sulfate-reducing bacteria.

Controlled Permeability Formwork

Controlled permeability formwork (CPF) is a system proven to significantly enhance the durability of surface concrete during the casting process.

The Need for CPF

All concretes cast against wood or steel in the coverzone, have a reduced cement content and increased water/cement ratio (i.e. less dense and more porous) compared to concrete located beyond the reinforcement. Within the core of any structural element the matrix is generally a denser and of better quality compared to the surface as a direct result of the concrete compaction. This compaction process drives excess mix air and water within the coverzone towards the formed surfaces.

As conventional wood or steel formwork is impermeable (IMF), the migration within the mix ceases as the concrete/formwork interface is reached. Visually, this may be observed on all concrete surfaces through the presence of blowholes following formwork removal.

This is a problem because the first line of defence of all structural elements against carbonation, chlorides, frost and abrasion, is the coverzone. It is therefore imperative this region is durable.

CPF Liners

CPF liners are typically constructed of 100% polypropylene fibres, spun and thermally bonded, with a woven texture of 0.7 mm thickness. Some systems may be laminated to a plastic latticed net to ensure drainage whilst providing stiffness to the liner.

They are robust and chemically inert and may be supplied in rolls of varying length to construction sites.

Basic Elements of CPF Systems

The basic elements of CPF systems are:

- *A filter*-to allow the passage of water and air away from the fresh concrete but retain cement and other fine solids.
- *A drain*-which transfers air and water from the filter to outside the formwork.
- *A structural support*-this is the formwork which supports the filter and drainage elements and also maintains the required formwork profile and resists the concrete pressure, commonly plywood/structural timber.

How They Work

CPF liners are secured in place on vertical or inclined surfaces with staples or other fixing devices, having first been tensioned onto the formwork shuttering. Once attached concreting is performed in the normal way. Release agents are not required as CPF liners easily debond from the concrete during formwork striking.

Throughout the concreting process and as a result of concrete pressures, entrapped air and excess mix water that would otherwise become trapped at the surface causing blemishes, can instead pass through the liner. A proportion of this mix water is held within the liner and under capillary action, imbibes back into the concrete to assist curing. Liners generally have a pore structure that is designed to retain the majority of cement and other small fines.

This results (for vertical and inclined surfaces) in the creation a uniform surface relatively free from blowholes and other surface blemishes when compared to IMF concrete. But more importantly the achievement of a cover area with significantly enhanced durability.

Cathodic Protection

Cathodic protection (CP) is a technique used to control the corrosion of a metal surface by making it the cathode of an electrochemical cell. The simplest method to apply CP is by connecting the metal to be protected with another more easily corroded "sacrificial metal" to act as the anode of the electrochemical cell. Cathodic protection systems are used to protect a wide range of metallic structures in various environments. Common applications are; steel water or fuel pipelines and storage tanks; steel pier piles; ships and boats; offshore oil platforms and onshore oil well casings and metal reinforcement bars in concrete buildings and structures.

Cathodic protection can, in some cases, prevent stress corrosion cracking.

History

Cathodic protection was first described by Sir Humphry Davy in a series of papers presented to the Royal Society in London in 1824. After a series of tests, the first application was to the HMS Samarang in 1824. Sacrificial anodes made from iron were attached to the copper sheath of the hull below the waterline and dramatically reduced the corrosion rate of the copper. However, a side effect of the CP was to increase marine growth. Copper, when corroding, releases copper ions which have an anti-fouling effect. Since excess marine growth affected the performance of the ship, the Royal Navy decided that it was better to allow the copper to corrode and have the benefit of reduced marine growth, so CP was not used further.

Davy was assisted in his experiments by his pupil Michael Faraday, who continued his research after Davy's death. In 1834, he discovered the quantitative connection between corrosion weight loss and electric current and thus laid the foundation for the future application of cathodic protection.

Thomas Edison experimented with impressed current cathodic protection on ships in 1890, but was unsuccessful due to the lack of a suitable current source and anode materials. It would be 100 years after Davy's experiment before cathodic protection was used widely on oil pipelines in the United States

Types

Galvanic CP

Galvanic anodes are designed and selected to have a more "active" voltage (more negative electrochemical potential) than the metal of the structure (typically steel). For effective CP, the potential of the steel surface is polarized (pushed) more negative until the surface has a uniform potential. At that stage, the driving force for the corrosion reaction is removed. The galvanic anode continues to corrode, consuming the anode material until eventually it must be replaced. The polarization is caused by the electron flow from the anode to the cathode. The driving force for the CP current is the difference in electrochemical potential between the anode and the cathode.

Galvanic or sacrificial anodes are made in various shapes and sizes using alloys of zinc, magnesium and aluminum. ASTM International publishes standards on the composition and manufacturing of galvanic anodes.

In order for galvanic cathodic protection to work, the anode must possess a lower (that is, more negative) potential than that of the cathode (the structure to be protected). The table below shows a simplified galvanic series to show which metals can thus be combined.

Metal	***Potential with Respect to a Cu:CuSO$_4$ Reference electrode in neutral pH environment (volts)***
Carbon, Graphite, Coke	+0.3
Platinum	0 to-0.1
Mill scale on Steel	-0.2
High Silicon Cast Iron	-0.2
Copper, brass, bronze	-0.2
Mild steel in concrete	-0.2
Lead	-0.5
Cast iron (not graphitized)	-0.5
Mild steel (rusted)	-0.2 to -0.5
Mild steel (clean)	-0.5 to -0.8
Commercially pure aluminum	-0.8
Aluminum alloy (5% zinc)	-1.05
Zinc	-1.1
Magnesium Alloy (6% Al, 3% Zn, 0.15% Mn)	-1.6
Commercially Pure Magnesium	-1.75

Impressed Current CP

For larger structures, galvanic anodes cannot deliver economically enough current to provide complete protection. Impressed current cathodic protection (ICCP) systems use anodes connected to a DC power source. Usually this will be a cathodic protection rectifier, which converts an AC power supply to a DC output. In the absence of an AC supply, alternative power sources may be used, such as solar panels, wind power or gas powered thermoelectric generators

Anodes for ICCP systems are available in a variety of shapes and sizes. Common anodes are tubular and solid rod shapes or continuous ribbons of various materials. These include high silicon cast iron, graphite, mixed metal oxide, platinum and niobium coated wire and others.

Applications

Pipelines

Pipelines are routinely protected by a coating supplemented with cathodic protection. An ICCP system for a pipeline would consist of a DC power source, which is often an AC powered rectifier and an anode, or array of anodes buried in the ground (the anode groundbed).

The DC power source would typically have a DC output of between 10 and 50 amperes and 50 volts, but this depends on several factors, such as the size of the pipeline. The positive DC output terminal would be connected via cables to the anode array, while another cable would connect the negative terminal of the rectifier to the pipeline, preferably through junction boxes to allow measurements to be taken.

Anodes can be installed in a vertical hole and backfilled with conductive coke (a material that improves the performance and life of the anodes) or laid in a prepared trench, surrounded by conductive coke and backfilled. The choice of grounded type and size depends on the application, location and soil resistivity.

The output of the DC source would then be adjusted to the optimum level after conducting various tests including measurements of electrochemical potential.

It is sometimes more economically viable to protect a pipeline using galvanic anodes. This is often the case on smaller diameter pipelines of limited length.

Ships

Cathodic protection on ships is often implemented by galvanic anodes attached to the hull, rather than using ICCP. Since ships are regularly removed from the water for inspections and maintenance, it is a simple task to replace the galvanic anodes. Galvanic anodes are generally shaped to reduced drag in the water and fitted flush to the hull to also try to minimize drag.

Smaller vessels, with non-metallic hulls, such as yachts, will also use galvanic anodes to protect areas such as the rudder, but depend on an electrical connection between the anode and the item to be protected.

For ICCP on ships, a DC power supply is provided within the ship and the anodes mounted on the outside of the hull. The anode cables are introduced into the ship via a compression seal fitting and routed to the DC power source. The negative cable from the power supply is simply attached to the hull to complete the circuit. Ship ICCP anodes are flush-mounted, minimizing the effects of drag on the ship, and located a minimum 5 ft below the light load line in an area to avoid mechanical damage. The current density required for protection is a function of velocity and considered when selecting the current capacity and location of anode placement on the hull.

Some ships may require specialist treatment, for example aluminum hulls with steel fixtures will create an electrochemical cell where the aluminum hull can act as a galvanic anode and corrosion is enhanced. In cases like this, aluminum or zinc galvanic anodes can be used to offset the potential difference between the aluminum hull and the steel fixture. If the steel fixtures are large, several galvanic anodes may be required, or even a small ICCP system.

Marine

Marine CP covers many areas, jetties, harbours, offshore structures. The variety of different types of structure leads to a variety of systems to provide protection. Typically, galvanic anodes are favoured, but ICCP can also often be used.

Steel in Concrete

The application to concrete reinforcement is slightly different in that the anodes and reference electrodes are usually embedded in the concrete at the time of construction when the concrete is being poured. The usual technique for concrete buildings, bridges and similar structures is to use ICCP, but there are systems available that use the principle of galvanic CP as well, although in the UK at least, the use of galvanic anodes for atmospherically exposed reinforced concrete structures is considered experimental.

For ICCP, the principle is the same as any other ICCP system. However, in a typical atmospherically exposed concrete structure such as a bridge, there will be many more anodes distributed through the structure as opposed to an array of anodes as used on a pipeline. This makes for a more complicated system and usually an automatically controlled DC power source is used, possibly with an option for remote monitoring and operation. For buried or submerged structures, the treatment is similar to that of any other buried or submerged structure.

Galvanic systems offer the advantage of being easier to fit retrospectively, since the anodes are fitted on the concrete surface and do not need any control systems as ICCP does.

For pipelines constructed from pre-stressed concrete cylinder pipe (PCCP), the techniques used for CP are generally as for steel pipelines except that there is a need to take steps to avoid an excessive level of potential that can produce cause possible damage to the prestressing wire

The steel wire in a PCCP pipeline is stressed to the point that any corrosion of the wire can result in failure. An additional problem is that any excessive hydrogen ions as a result of an excessively negative potential can cause hydrogen embrittlement of the wire, also resulting in failure. The failure of too many wires will result in catastrophic failure of the PCCP. To implement ICCP therefore requires very careful control to ensure satisfactory protection. A simpler option is to use galvanic anodes, which are self limiting and need no control.

Internal CP

Vessels, pipelines and tanks which are used to store or transport liquids can also be protected from corrosion on their internal surfaces by the use of cathodic protection. ICCP and galvanic systems can be used.

Galvanized Steel

Galvanizing generally refers to hot-dip galvanizing which is a way of coating steel with a layer of metallic zinc. Galvanized coatings are quite durable in most environments because they combine the barrier properties of a coating with some of the benefits of cathodic protection. If the zinc coating is scratched or otherwise locally damaged and steel is exposed, the surrounding areas of zinc coating form a galvanic cell with the exposed steel and protect it from corrosion. This is a form of localized cathodic protection-the zinc acts as a sacrificial anode. It should be noted that galvanizing, while using the principle of cathodic protection, is not actually cathodic protection. CP requires the anode to be separate from the metal surface to be protected, with an ionic connection through the electrolyte and an electron connection through a connecting cable, bolt or similar. This means that any area of the protected structure within the electrolyte can be protected, whereas in the case of galvanizing, only areas very close to the zinc are protected. Hence, a larger area of bare steel would only be protected around the edges.

Testing

Electrochemical corrosion potential is measured with reference electrodes. Copper-copper(II) sulfate electrodes are used for structures in contact with soil or fresh water. Silver chloride electrodes or Saturated calomel electrodes (SCE) are used for seawater applications. The methods are described in along with the sources of error in the voltage that appears on the display of the meter. Interpretation of electrochemical corrosion potential measurements to determine the potential at the interface between the anode of the corrosion cell and the electrolyte requires training and cannot be expected to match the accuracy of measurements done in laboratory work. It is widely recognized that errors can be introduced in all measurements where there is a lack of training.

Problems

Production of Hydrogen ions

A side effect of improperly applied cathodic protection is the production of hydrogen ions, leading to its absorption in the protected metal and subsequent hydrogen embrittlement of welds and materials with high hardness. Under normal conditions, the ionic hydrogen will combine at the metal surface to create hydrogen gas, which cannot penetrate the metal. Hydrogen ions, however, are small enough to pass through the crystalline steel structure, and lead in some cases to hydrogen embrittlement.

Cathodic Disbonding

This is a process of disbondment of protective coatings from the protected structure (cathode) due to the formation of hydrogen ions over the surface of the protected material (cathode). Disbonding can be exacerbated by an increase in alkali ions and an increase in cathodic polarization. The degree of disbonding is also reliant on the type of coating, with some coatings affected more than others. Cathodic protection systems should be operated so that the structure does not become excessively polarized, since this also promotes disbonding due to excessively negative potentials. Cathodic disbonding occurs rapidly in pipelines that contain hot fluids because the process is accelerated by heat flow.

Cathodic Shielding

Effectiveness of cathodic protection systems on steel pipelines can be impaired by the use of solid film backed dielectric coatings such as polyethylene tapes, shrinkable pipeline sleeves, and factory applied single or multiple solid film coatings. This phenomenon occurs because of the high electrical resistivity of these film backings. Protective electric current from the cathodic protection system is blocked or shielded from reaching the underlying metal by the highly resistive film backing. Cathodic shielding was first defined in the 1980s as being a problem, and technical papers on the subject have been regularly published since then.

A 1999 report concerning a 20,600 bbl (3,280 m^3). spill from a Saskatchewan crude oil line contains an excellent definition of the cathodic shielding problem:

"The triple situation of disbondment of the (corrosion) coating, the dielectric nature of the coating and the unique electrochemical environment established under the exterior coating, which acts as a shield to the electrical CP current, is referred to as CP shielding. The combination of tenting and disbondment permits a corrosive environment around the outside of the pipe to enter into the void between the exterior coating and the pipe surface. With the development of this CP shielding phenomenon, impressed current from the CP system cannot access exposed metal under the exterior coating to protect the pipe surface from the consequences of an aggressive corrosive environment.

The CP shielding phenomenon induces changes in the potential gradient of the CP system across the exterior coating, which are further pronounced in areas of insufficient or sub-standard CP current emanating from the pipeline's CP system. This produces an area on the pipeline of insufficient CP defence against metal loss aggravated by an exterior corrosive environment."

Cathodic shielding is referenced in a number of the standards listed below. Newly issued USDOT regulation Title 49 CFR 192.112, in the section for *Additional design requirements for steel pipe using alternative maximum allowable operating pressure* requires that "The pipe must be protected against external corrosion by a non-shielding coating". Also, the NACE SP0169:2007 standard defines shielding in section 2, cautions against the use of materials that create electrical shielding in section 4.2.3, cautions against use of external coatings that create electrical shielding in section 5.1.2.3, and instructs readers to take 'appropriate action' when the effects of electrical shielding of CP current are detected on an operating pipeline in section 10.9.

Standards

- 49 CFR 192.112-Requirements for Corrosion Control-Transportation of natural and other gas by pipeline: minimum federal safety standards
- ASME B31Q 0001-0191
- ASTM G 8, G 42-Evaluating Cathodic Disbondment resistance of coatings
- DNV-RP-B401-Cathodic Protection Design-Det Norske Veritas
- EN 12068:1999-Cathodic protection. External organic coatings for the corrosion protection of buried or immersed steel pipelines used in conjunction with cathodic protection. Tapes and shrinkable materials
- EN 12473:2000-General principles of cathodic protection in sea water
- EN 12474:2001-Cathodic protection for submarine pipelines
- EN 12495:2000-Cathodic protection for fixed steel offshore structures
- EN 12499:2003-Internal cathodic protection of metallic structures
- EN 12696:2000-Cathodic protection of steel in concrete
- EN 12954:2001-Cathodic protection of buried or immersed metallic structures. General principles and application for pipelines
- EN 13173:2001-Cathodic protection for steel offshore floating structures
- EN 13174:2001-Cathodic protection for harbour installations
- EN 13509:2003-Cathodic protection measurement techniques

- EN 13636:2004-Cathodic protection of buried metallic tanks and related piping
- EN 14505:2005-Cathodic protection of complex structures
- EN 15112:2006-External cathodic protection of well casing
- EN 50162:2004-Protection against corrosion by stray current from direct current systems
- BS 7361-1:1991-Cathodic Protection
- NACE SP0169:2007-Control of External Corrosion on Underground or Submerged Metallic Piping Systems
- NACE TM 0497-Measurement Techniques Related to Criteria for Cathodic Protection on Underground or Submerged Metallic Piping Systems.

Sacrificial Anode Protection

For effective CP, the potential of the steel surface is polarized (pushed) more negative until the metal surface has a uniform potential. With a uniform potential, the driving force for the corrosion reaction is halted. For galvanic CP systems, the anode material corrodes under the influence of the steel, and eventually it must be replaced. The polarization is caused by the current flow from the anode to the cathode, driven by the difference in electrochemical potential between the anode and the cathode.

Impressed Current Cathodic Protection

For larger structures, galvanic anodes cannot economically deliver enough current to provide complete protection. Impressed Current Cathodic Protection (ICCP) systems use anodes connected to a DC power source (such as a cathodic protection rectifier). Anodes for ICCP systems are tubular and solid rod shapes of various specialized materials. These include high silicon cast iron, graphite, mixed metal oxide or platinum coated titanium or niobium coated rod and wires.

Anodic Protection

Anodic protection (AP) is a technique to control the corrosion of a metal surface by making it the anode of an electrochemical cell and controlling the electrode potential in a zone where the metal is passive.

AP is used to protect metals that exhibit passivation in environments whereby the current density in the freely corroding state is significantly higher than the current density in the passive state over a wide range of potentials. Anodic protection is used for carbon steel storage tanks containing extreme pH environments including concentrated sulfuric acid and 50 percent caustic soda where cathodic protection is not suitable due to very high current requirements.

An anodic protection system includes an external power supply connected to auxiliary cathodes and controlled by a feedback from reference electrode. Careful design and control is required when using anodic protection for several reasons, including excessive current when passivation is lost or unstable, leading to possible accelerated corrosion.

Economic Impact

The US Federal Highway Administration released a study, entitled *Corrosion Costs and Preventive Strategies in the United States,* in 2002 on the direct costs associated with metallic corrosion in nearly every U.S. industry sector. The study showed that for 1998 the total annual estimated direct cost of corrosion in the U.S. was approximately \$276 billion (approximately 3.2% of the US gross domestic product).

Rust is one of the most common causes of bridge accidents. As rust has a much higher volume than the originating mass of iron, its build-up can also cause failure by forcing apart adjacent parts. It was the cause of the collapse of the Mianus river bridge in 1983, when the bearings rusted internally and pushed one corner of the road slab off its support.

Three drivers on the roadway at the time died as the slab fell into the river below. The following NTSB investigation showed that a drain in the road had been blocked for road re-surfacing, and had not been unblocked so that runoff water penetrated the support hangers. It was also difficult for maintenance engineers to see the bearings from the inspection walkway. Rust was also an important factor in the Silver Bridge disaster of 1967 in West Virginia, when a steel suspension bridge collapsed in less than a minute, killing 46 drivers and passengers on the bridge at the time.

Similarly, corrosion of concrete-covered steel and iron can cause the concrete to spall, creating severe structural problems. It is one of the most common failure modes of reinforced concrete bridges. Measuring instruments based on the half-cell potential are able to detect the potential corrosion spots before total failure of the concrete structure is reached.

Corrosion in Nonmetals

Most ceramic materials are almost entirely immune to corrosion. The strong ionic and/or covalent bonds that hold them together leave very little free chemical energy in the structure; they can be thought of as already corroded. When corrosion does occur, it is almost always a simple dissolution of the material or chemical reaction, rather than an electrochemical process. A common example of corrosion protection in ceramics is the lime added to soda-lime glass to reduce its solubility in water; though it is not nearly as soluble as pure sodium silicate, normal glass does form sub-microscopic flaws when exposed to moisture. Due to its brittleness, such flaws cause a dramatic reduction in the strength of a glass object during its first few hours at room temperature.

Corrosion of Polymers

Polymer degradation is due to a wide array of complex and often poorly-understood physiochemical processes. These are strikingly different from the other processes discussed here, and so the term "corrosion" is only applied to them in a loose sense of the word. Because of their large molecular weight, very little entropy can be gained by mixing a given mass of polymer with another substance, making them generally quite difficult to dissolve. While dissolution is a problem in some polymer applications, it is relatively simple to design against. A more common and related problem is *swelling*, where small molecules infiltrate the structure, reducing strength and stiffness and causing a volume change. Conversely, many polymers (notably flexible vinyl) are intentionally swelled with plasticizers, which can be leached out of the structure, causing brittleness or other undesirable changes. The most common form of degradation, however, is a decrease in polymer chain length. Mechanisms which break polymer chains are familiar to biologists because of their effect on DNA: ionizing radiation (most commonly ultraviolet light), free radicals, and oxidizers such as oxygen, ozone, and chlorine. Ozone cracking is a well-known problem affecting natural rubber for example. Additives can slow these process very effectively, and can be as simple as a UV-absorbing pigment (i.e., titanium dioxide or carbon black). Plastic shopping bags often do not include these additives so that they break down more easily as litter.

Corrosion of Glasses

Glass disease is the corrosion of silicate glasses in aqueous solutions. It is governed by two mechanisms: diffusion-controlled leaching (ion exchange) and glass network hydrolytic dissolution. Both corrosion mechanisms strongly depend on the pH of contacting solution: the rate of ion exchange decreases with pH as $10^{-0.5pH}$ whereas the rate of hydrolytic dissolution increases with pH as $10^{0.5pH}$

Mathematically, corrosion rates of glasses are characterised by normalised corrosion rates of elements NR_i (g/cm^2 d) which are determined as the ratio of total amount of released species into the water M_i (g) to the water-contacting surface area S (cm^2), time of contact t (days) and weight fraction content of the element in the glass f_i:

$$NR_i = \frac{M_i}{Sf_i t}.$$

The overall corrosion rate is a sum of contributions from both mechanisms (leaching + dissolution) $NR_i = Nrx_i + NRh$. Diffusion-controlled leaching (ion exchange) is characteristic of the initial phase of corrosion and involves replacement of alkali ions in the glass by a hydronium (H_3O^+) ion from the solution.

It causes an ion-selective depletion of near surface layers of glasses and gives an inverse square root dependence of corrosion rate with exposure time. The diffusion controlled normalised leaching rate of cations from glasses (g/cm^2 d) is given by:

$$NRx_i = 2\rho\sqrt{\frac{D_i}{\pi t}},$$

where t is time, D_i is the i-th cation effective diffusion coefficient (cm^2/d), which depends on pH of contacting water as $D_i = D_{i0} \cdot 10^{-pH}$, and ρ is the density of the glass (g/cm^3).

Glass network dissolution is characteristic of the later phases of corrosion and causes a congruent release of ions into the water solution at a time-independent rate in dilute solutions (g/cm^2 d):

$$NRh = \rho r_h,$$

where r_h is the stationary hydrolysis (dissolution) rate of the glass (cm/d). In closed systems the consumption of protons from the aqueous phase increases the pH and causes a fast transition to hydrolysis. However further silica saturation of solution impedes hydrolysis and causes the glass to return to an ion-exchange, e.g. diffusion-controlled regime of corrosion.

In typical natural conditions normalised corrosion rates of silicate glasses are very low and are of the order of 10^{-7}-10^{-5} g/cm^2 d. The very high durability of silicate glasses in water makes them suitable for hazardous and nuclear waste immobilisation.

Glass Corrosion Tests

There exist numerous standardized procedures for measuring the corrosion (also called chemical durability) of glasses in neutral, basic, and acidic environments, under simulated environmental conditions, in simulated body fluid, at high temperature and pressure, and under other conditions.

In the standard procedure ISO 719 a test of the extraction of water soluble basic compounds under neutral conditions is described: 2 g glass, particle size 300-500 μm, is kept for 60 min in 50 ml de-ionized water of grade 2 at 98°C. 25 ml of the obtained solution is titrated against 0.01 mol/l HCl solution. The volume of HCl needed for neutralization is recorded and classified following the values in the table below.

Heat Treating

Heat treating is a group of industrial and metalworking processes used to alter the physical, and sometimes chemical, properties of a material. The most common application is metallurgical. Heat treatments are also used in the manufacture of many other materials, such as glass. Heat treatment involves the use of heating or chilling, normally to extreme temperatures, to achieve a desired result such as hardening or softening of a material.

Heat treatment techniques include annealing, case hardening, precipitation strengthening, tempering and quenching. It is noteworthy that while the term *heat treatment* applies only to processes where the heating and cooling are done for the specific purpose of altering properties intentionally, heating and cooling often occur incidentally during other manufacturing processes such as hot forming or welding.

Physical Processes

Metallic materials consist of a microstructure of small crystals called "grains" or crystallites. The nature of the grains (i.e. grain size and composition) is one of the most effective factors that can determine the overall mechanical behaviour of the metal. Heat treatment provides an efficient way to manipulate the properties of the metal by controlling the rate of diffusion and the rate of cooling within the microstructure. There are two mechanisms that may change an alloy's properties during heat treatment. The martensite transformation causes the crystals to deform intrinsically. The diffusion mechanism causes changes in the homogeneity of the alloy.

The crystal structure consists of atoms that are grouped in a very specific arrangement, called a lattice. In most elements, this order will rearrange itself, depending on conditions like temperature and pressure. This rearrangement, called allotropy or polymorphism, may occur several times, at many different temperatures for a particular metal. In alloys, this rearrangement may cause an element that will not normally dissolve into the base metal to suddenly become soluble, while a reversal of the allotropy will make the elements either partially or completely insoluble.

When in the soluble state, the process of diffusion causes the atoms of the dissolved element to spread out, attempting to form a homogenous distribution within the crystals of the base metal. If the alloy is cooled to an insoluble state, the atoms of the dissolved constituents (solutes) may migrate out of the solution. This type of diffusion, called precipitation, leads to nucleation, where the migrating atoms group together at the grain-boundaries. This forms a microstructure generally consisting of two or more distinct phases. Steel that has been cooled slowly, for instance, forms a laminated structure composed of alternating layers of ferrite and cementite, becoming soft pearlite.

Unlike iron-based alloys, most heat treatable alloys do not experience a ferrite transformation. In these alloys, the nucleation at the grain-boundaries often reinforces the structure of the crystal matrix. These metals harden by preciptation. Typically a slow process, depending on temperature, this is often referred to as "age hardening".

Many metals and non-metals exhibit a martensite transformation when cooled quickly. When a metal is cooled very quickly, the insoluble atoms may not be able to migrate out of the solution in time. This is called a "diffusionless transformation." When the crystal matrix changes to its low temperature arrangement, the atoms of the solute become trapped within the lattice.

The trapped atoms prevent the crystal matrix from completely changing into its low temperature allotrope, creating shearing stresses within the lattice. When some alloys are cooled quickly, such as steel, the martensite transformation hardens the metal, while in others, like aluminum, the alloy becomes softer.

Effects of Composition

The specific composition of an alloy system will usually have a great effect on the results of heat treating. If the percentage of each constituent is just right, the alloy will form a single, continuous microstructure upon cooling. Such a mixture is said to be eutectoid. However, If the percentage of the solutes varies from the eutectoid mixture, two or more different microstructures will usually form simultaneously. A hypoeutectoid solution contains less of the solute than the eutectoid mix, while a hypereutectoid solution contains more.

Eutectoid Alloys

A eutectoid alloy is similar in behaviour to a eutectic alloy. A *eutectic* alloy is characterized by having a single melting point. This melting point is lower than that of any of the constituents, and no change in the mixture will lower the melting point any further. When a molten eutectic alloy is cooled, all of the constituents will crystallize into their respective phases at the same temperature.

A eutectoid alloy is similar, but the phase change occurs, not from a liquid, but from a solid solution. Upon cooling a eutectoid alloy from the solution temperature, the constituents will separate into different crystal phases, forming a single microstructure. A eutectoid steel, for example, contains 0.77% carbon. Upon cooling slowly, the solution of iron and carbon, (a single phase called austenite), will separate into platelets of the phases ferrite and cementite. This forms a layered microstructure called pearlite.

Since pearlite is harder than iron, the degree of softness achieveable is typically limited to that produced by the pearlite. Similarly, the hardenability is limited by the continuous martensitic microstructure formed when cooled very fast.

Hypoeutectoid Alloys

A *hypoeutectic* alloy has two separate melting points. Both are above the eutectic melting point for the system, but are below the melting points of any constituent forming the system. Between these two melting points, the alloy will exist as part solid and part liquid. The constituent with the lower melting point will solidify first. When completely solidified, a hypoeutectic alloy will often be in solid solution.

Similarly, a hypoeutectoid alloy has two critical temperatures, called "arrests." Between these two temperatures, the alloy will exist partly as the solution and partly as a separate crystallizing phase. These two temperatures are called the upper (A_3) and lower (A_1) transformation temperatures. As the solution cools from the upper transformation temperature toward an insoluble state, the excess base metal will often be forced to "crystallize-out." This will occur until the remaining concentration of solutes reaches the eutectoid level, which will then crystallize as a separate microstructure.

A hypoeutectoid steel contains less than 0.77% carbon. Upon cooling a hypoeutectoid steel from the austenite transformation temperature, small islands of ferrite will form. These will continue to grow until the eutectoid concentration in the rest of the steel is reached. This eutectoid mixture will then crystallize as a microstructure of pearlite. Since ferrite is softer than pearlite, the two microstructures combine to increase the ductility of the alloy. Consequently, the hardenability of the alloy is lowered.

Hypereutectoid Alloys

A *hypereutectic* alloy also has different melting points. However, between these points, it is the constituent with the higher melting point that will be solid. Similarly, a hypereutectoid alloy has two critical temperatures. When cooling a hypereutectoid alloy from the upper transformation temperature, it will usually be the excess solutes that crystallize-out first. This continues until the concentration in the remaining alloy becomes eutectoid, which then crystallizes into a separate microstructure.

A hypereutectoid steel contains more than 0.77% carbon. When slowly cooling a hypereutectoid steel, the cementite will begin to crystallize first. When the remaining steel becomes eutectoid in composition, it will crystallize into pearlite. Since cementite is much harder than pearlite, the alloy has greater hardenability at a cost in the ductility.

Effects of Time and Temperature

Proper heat treating requires precise control over temperature, time held at a certain temperature and cooling rate.

With the exception of stress-relieving, tempering, and aging, most heat treatments begin by heating an alloy beyond the upper transformation (A_3) temperature. The alloy will usually be held at this temperature long enough for the heat to completely penetrate the alloy, thereby bringing it into a complete solid solution. Since a smaller grain size usually enhances mechanical properties, such as toughness, shear strength and tensile strength, these metals are often heated to a temperature that is just above the upper critical temperature, in order to prevent the grains of solution from growing too large. For instance, when steel is heated above the upper critical temperature, small grains of austenite form. These grow larger as temperature is increased. When cooled very quickly, during a martensite transformation, the austenite grain size directly affects the martensitic grain size. Larger grains have large grain-boundaries, which serve as weak spots in the structure. The grain size is usually controlled to reduce the probability of breakage.

The diffusion transformation is very time dependent. Cooling a metal will usually suppress the precipitation to a much lower temperature. Austenite, for example, usually only exists above the upper critical temperature. However, if the austenite is cooled quickly enough, the transformation may be suppressed for hundreds of degrees below the lower critical temperature. Such austenite is highly unstable and, if given enough time, will precipitate into various microstructures of ferrite and cementite. The cooling rate can be used to control the rate of grain growth or can even be used to produce partially martensitic microstructures. However, the martensite transformation is time-independent. If the alloy is cooled to the martensite transformation (M_s) temperature before other microstructures can fully form, the transformation will usually occur at just under the speed of sound.

When austenite is cooled slow enough that a martensite transformation does not occur, the austenite grain size will have an effect on the rate of nucleation, but it is generally temperature and the rate of cooling that controls the grain size and microstructure. When austenite is cooled extremely slow, it will form large ferrite crystals filled with spherical inclusions of cementite. This microstructure is referred to as "sphereoidite." If cooled a little faster, then coarse pearlite will form. Even faster, and fine pearlite will form. If cooled even faster, bainite will form. Similarly, these microstructures will also form if cooled to a specific temperature and then held there for a certain time.

Most non-ferrous alloys are also heated in order to form a solution. Most often, these are then cooled very quickly to produce a martensite transformation, putting the solution into a supersaturated state. The alloy, being in a much softer state, may then be cold worked. This cold working increases the strength and hardness of the alloy, and the defects caused by plastic deformation tend to speed up precipitation, increasing the hardness beyond what is normal for the alloy. Even if not cold worked, the solutes in these alloys will usually precipitate, although the process may take much longer. Sometimes these metals are then heated to a temperature that is below the lower critical (A_1) temperature, preventing recrystallization, in order to speed-up the precipitation.

Techniques

Complex heat treating schedules, or "cycles," are often devised by metallurgists to optimize an alloy's mechanical properties. In the aerospace industry, a superalloy may undergo five or more different heat treating operations to develop the desired properties. This can lead to quality problems depending on the accuracy of the furnace's temperature controls and timer.

Annealing (metallurgy)

Annealing, in metallurgy and materials science, is a heat treatment wherein a material is altered, causing changes in its properties such as strength and hardness. It is a process that produces conditions by heating to above the recrystallization temperature, maintaining a suitable temperature, and then cooling. Annealing is used to induce ductility, soften material, relieve internal stresses, refine the structure by making it homogeneous, and improve cold working properties.

In the cases of copper, steel, silver, and brass, this process is performed by substantially heating the material (generally until glowing) for a while and allowing it to cool. Unlike ferrous metals—which must be cooled slowly to anneal—copper, silver and brass can be cooled slowly in air or quickly by quenching in water. In this fashion the metal is softened and prepared for further work such as shaping, stamping, or forming.

Thermodynamics

Annealing occurs by the diffusion of atoms within a solid material, so that the material progresses towards its equilibrium state. Heat is needed to increase the rate of diffusion by providing the energy needed to break bonds. The movement of atoms has the effect of redistributing and destroying the dislocations in metals and (to a lesser extent) in ceramics. This alteration in dislocations allows metals to deform more easily, so increases their ductility. The amount of process-initiating Gibbs free energy in a deformed metal is also reduced by the annealing process. In practice and industry, this reduction of Gibbs free energy is termed "stress relief".

The relief of internal stresses is a thermodynamically spontaneous process; however, at room temperatures, it is a very slow process. The high temperatures at which the annealing process occurs serve to accelerate this process.

The reaction facilitating the return of the cold-worked metal to its stress-free state has many reaction pathways, mostly involving the elimination of lattice vacancy gradients within the body of the metal. The creation of lattice vacancies is governed by the Arrhenius equation, and the migration/diffusion of lattice vacancies are governed by Fick's laws of diffusion.

Mechanical properties, such as hardness and ductility, change as dislocations are eliminated and the metal's crystal lattice is altered. On heating at specific temperature and cooling it is possible to bring the atom at the right lattice site and new grain growth can improve the mechanical properties.

Stages

There are three stages in the annealing process, with the first being the recovery phase, which results in softening of the metal through removal of crystal defects (the primary type of which is the linear defect called a dislocation) and the internal stresses which they cause. Recovery phase covers all annealing phenomena that occur before the appearance of new strain-free grains. The second phase is recrystallization, where new strain-free grains nucleate and grow to replace those deformed by internal stresses. If annealing is allowed to continue once recrystallization has been completed, grain growth will occur, in which the microstructure starts to coarsen and may cause the metal to have less than satisfactory mechanical properties.

Controlled Atmospheres

The high temperature of annealing may result in oxidation of the metal's surface, resulting in scale. If scale is to be avoided, annealing is carried out in a special atmosphere, such as with endothermic gas (a mixture of carbon monoxide, hydrogen gas, and nitrogen gas). Annealing is also done in forming gas, a mixture of hydrogen and nitrogen.

The magnetic properties of mu-metal (Espey cores) are introduced by annealing the alloy in a hydrogen atmosphere.

Setup and Equipment

Typically, large ovens are used for the annealing process. The inside of the oven is large enough to place the workpiece in a position to receive maximum exposure to the circulating heated air. For high volume process annealing, gas fired conveyor furnaces are often used. For large workpieces or high quantity parts Car-bottom furnaces will be used in order to move the parts in and out with ease. Once the annealing process has been successfully completed, the workpieces are sometimes left in the oven in order for the parts to have a controlled cooling process. While some workpieces are left in the oven to cool in a controlled fashion, other materials and alloys are removed from the oven. After being removed from the oven, the workpieces are often quickly cooled off in a process known as quench hardening. Some typical methods of quench hardening materials involve the use of media such as air, water, oil, or salt. Quench hardening is generally applicable to some ferrous alloys, but not copper alloys.

Diffusion Annealing of Semiconductors

In the semiconductor industry, silicon wafers are annealed, so that dopant atoms, usually boron, phosphorus or arsenic, can diffuse into substitutional positions in the crystal lattice, resulting in drastic changes in the electrical properties of the semiconducting material.

Specialized Cycles

Normalization

Normalization is an annealing process in which a metal is cooled in air after heating in order to relieve stress.

It can also be referred to as: Heating a ferrous alloy to a suitable temperature above the transformation temperature range and cooling in air to a temperature substantially below the transformation range.

This process is typically confined to hardenable steel. It is used to refine grains which have been deformed through cold work, and can improve ductility and toughness of the steel. It involves heating the steel to just above its upper critical point. It is soaked for a short period then allowed to cool in air. Small grains are formed which give a much harder and tougher metal with normal tensile strength and not the maximum ductility achieved by annealing. It eliminates columnar grains and dendritic segregation that sometimes occurs during casting. Normalizing improves machinability of a component and provides dimensional stability if subjected to further heat treatment processes.

Process Annealing

Process annealing, also called "intermediate annealing", "subcritical annealing", or "in-process annealing", is a heat treatment cycle that restores some of the ductility to a work piece allowing it be worked further without breaking. Ductility is important in shaping and creating a more refined piece of work through processes such as rolling, drawing, forging, spinning, extruding and heading. The piece is heated to a temperature typically below the austenizing temperature, and held there long enough to relieve stresses in the metal. The piece is finally cooled slowly to room temperature. It is then ready again for additional cold working. This can also be used to ensure there is reduced risk of distortion of the work piece during machining, welding, or further heat treatment cycles.

The temperature range for process annealing ranges from 260 °C(500 °F) to 760 °C(1400 °F), depending on the alloy in question.

Full Anneal

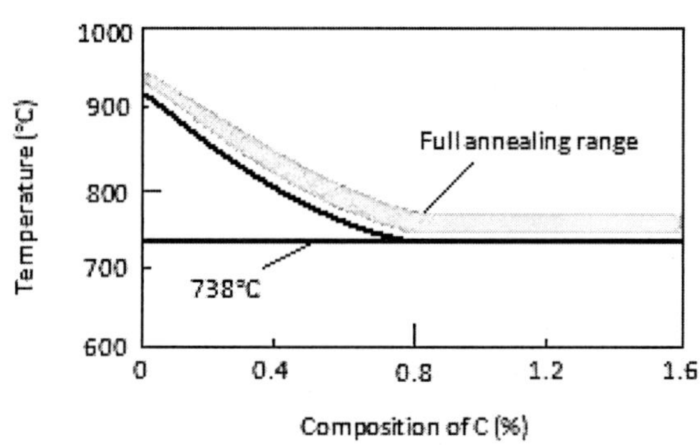

Table 1: Full Annealing Temperature Ranges

A full anneal typically results in the second most ductile state a metal can assume for metal alloy. It creates an entirely new homogeneous and uniform structure with good dynamic properties. To perform a full anneal on steel for example, steel is heated to its annealing point (about 50°C above the austenic temperature as graph shows) and held for sufficient time to allow the material to fully austenitize, to form austenite or austenite-cementite grain structure. The material is then allowed to cool slowly so that the equilibrium microstructure is obtained. In some cases this means the material is allowed to air cool. In other cases the material is allowed to furnace cool. The details of the process depend on the type of metal and the precise alloy involved. In any case the result is a more ductile material that has greater stretch ratio and reduction of area properties but a lower yield strength and a lower tensile strength. This process is also called LP annealing for *lamellar pearlite* in the steel industry as opposed to a *process anneal* which does not specify a microstructure and only has the goal of softening the material. Often material that is to be machined, will be annealed, then be followed by further heat treatment to obtain the final desired properties.

Short Cycle Anneal

Short cycle annealing is used for turning normal ferrite into malleable ferrite. It consists of heating, cooling, and then heating again from 4 to 8 hours.

Resistive Heating

Resistive heating can be used to efficiently anneal copper wire; the heating system employs a controlled electrical short circuit. It can be advantageous because it does not require a temperature-regulated furnace like other methods of annealing.

The process consists of two conductive pulleys (*step pulleys*) which the wire passes across after it is drawn. The two pulleys have an electrical potential across them, which causes the wire to form a short circuit. The Joule effect causes the temperature of the wire to rise to approximately 400 °C. This temperature is affected by the rotational speed of the pulleys, the ambient temperature, and the voltage applied. Where t is the temperature of the wire, K is a constant, V is the voltage applied, r is the number of rotations of the pulleys per minute, and t_a is the ambient temperature:

$$t = ((KV^2)/(r)) + t_a$$

The constant K depends on the diameter of the pulleys and the resistivity of the copper. Purely in terms of the temperature of the copper wire, an increase in the speed with which the wire passes through the pulley system has the same effect as an increase in resistance. Therefore, the speed with which the wire can be drawn through varies quadratically as the voltage applied.

Normalizing

Normalizing is a technique used to provide uniformity in grain size and composition throughout an alloy. The term is often used for ferrous alloys that have been heated above the upper critical temperature and then cooled in open air. Normalising gives harder and stronger steel, but with less ductile for same composition, than full annealing.

Stress Relieving

Stress relieving is a technique to remove or reduce the internal stresses created in a metal. These stresses may be caused in a number of ways, ranging from cold working to non-uniform cooling. Stress relieving is usually accomplished by heating a metal below the lower critical temperature and then cooling uniformly.

Precipitation Hardening

Precipitation hardening, also called age hardening, is a heat treatment technique used to increase the yield strength of malleable materials, including most structural alloys of aluminium, magnesium, nickel and titanium, and some stainless steels. It relies on changes in solid solubility with temperature to produce fine particles of an impurity phase, which impede the movement of dislocations, or defects in a crystal's lattice.

Since dislocations are often the dominant carriers of plasticity, this serves to harden the material. The impurities play the same role as the particle substances in particle-reinforced composite materials. Just as the formation of ice in air can produce clouds, snow, or hail, depending upon the thermal history of a given portion of the atmosphere, precipitation in solids can produce many different sizes of particles, which have radically different properties. Unlike ordinary tempering, alloys must be kept at elevated temperature for hours to allow precipitation to take place. This time delay is called aging. Solution treatment and aging is sometimes abbreviated "STA" in metals specs and certs.

Note that two different heat treatments involving precipitates can alter the strength of a material: solution heat treating and precipitation heat treating. Solid solution strengthening involves formation of a single-phase solid solution via quenching and leaves a material softer. Precipitation heat treating involves the addition of impurity particles to increase a material's strength.

Kinetics Versus Thermodynamics

This technique exploits the phenomenon of supersaturation, and involves careful balancing of the driving force for precipitation and the thermal activation energy available for both desirable and undesirable processes.

Nucleation occurs at a relatively high temperature (often just below the solubility limit) so that the kinetic barrier of surface energy can be more easily overcome and the maximum number of precipitate particles can form. These particles are then allowed to grow at lower temperature in a process called *aging*. This is carried out under conditions of low solubility so that thermodynamics drive a greater total volume of precipitate formation.

Diffusion's exponential dependence upon temperature makes precipitation strengthening, like all heat treatments, a fairly delicate process. Too little diffusion (*under aging*), and the particles will be too small to impede dislocations effectively; too much (*over aging*), and they will be too large and dispersed to interact with the majority of dislocations.

Alloy Design

Precipitation strengthening is possible if the line of solid solubility slopes strongly toward the centre of a phase diagram. While a large volume of precipitate particles is desirable, a small enough amount of the alloying element should be added that it remains easily soluble at some reasonable annealing temperature.

Elements used for precipitation strengthening in typical aluminum and titanium alloys, make up about 10% of their composition. While binary alloys are more easily understood as an academic exercise, commercial alloys often use three components for precipitation strengthening, in compositions such as Al(Mg, Cu) and Ti(Al, V). A large number of other constituents may be unintentional, but benign, or may be added for other purposes such as grain refinement or corrosion resistance. In some cases, such as many aluminum alloys, an increase in strength is achieved at the expense of corrosion resistance.

The addition of large amounts of nickel and chromium needed for corrosion resistance in stainless steels means that traditional hardening and tempering methods are not effective. However, precipitates of chromium, copper or other elements can strengthen the steel by similar amounts in comparison to hardening and tempering. The strength can be tailored by adjusting the annealing process, with lower initial temperatures resulting in higher strengths. The lower initial temperature increase driving force of nucleation. More driving force means more nucleation sites, and more sites, means more places for dislocations to be disrupted while the finished part is in use.

Many alloy systems allow the aging temperature to be adjusted. For instance, some aluminium alloys used to make rivets for aircraft construction are kept in dry ice from their initial heat treatment until they are installed in the structure. After this type of rivet is deformed into its final shape, aging occurs at room temperature and increases its strength, locking the structure together. Higher aging temperatures would risk over-aging other parts of the structure, and require expensive post-assembly heat treatment. Too high of an aging temperature promotes the precipitate to grow too readily.

Theory

The primary species of precipitation strengthening are second phase particles. These particles impede the movement of dislocations throughout the lattice. You can determine whether or not second phase particles will precipitate into solution from the solidus line on the phase diagram for the particles. Physically, this strengthening effect can be attributed both to size and modulus effects, and to interfacial or surface energy.

The presence of second phase particles often causes lattice distortions. These lattice distortions result when the precipitate particles differ in size and crystallographic structure from the host atoms. Smaller precipitate particles in a host lattice leads to a tensile stress, whereas larger precipitate particles leads to a compressive stress. Dislocation defects also create a stress field. Above the dislocation there is a compressive stress and below there is a tensile stress. Consequently, there is a negative interaction energy between a dislocation and a precipitate that each respectively cause a compressive and a tensile stress or vice versa. In other words, the dislocation will be attracted to the precipitate. In addition, there is a positive interaction energy between a dislocation and a precipitate that have the same type of stress field. This means that the dislocation will be repulsed by the precipitate.

Precipitate particles also serve by locally changing the stiffness of a material. Dislocations are repulsed by regions of higher stiffness.

Conversely, if the precipitate causes the material to be locally more compliant, then the dislocation will be attracted to that region.

Furthermore, a dislocation may cut through a precipitate particle. This interaction causes an increase in the surface area of the particle. The area created is

$$A = 2rb\pi$$

where, r is the radius of the particle and b is the magnitude of the burgers vector. The resulting increase in surface energy is

$$A = 2rb\pi$$

where γ_s is the surface energy. The dislocation can also bow around a precipitate particle.

Governing Equations

There are two equations to describe the two mechanisms for precipitation hardening:

Dislocations cutting through particles:

$$\tau = \frac{r\gamma\pi}{bL}$$

where τ is material strength, r is the second phase particle radius, γ is the surface energy, b is the magnitude of the Burgers vector, and L is the spacing between pinning points. This governing equation shows that the strength is proportional to r, the radius of the precipitate particles. This means that it is easier for dislocations to cut through a material with smaller second phase particles (small r). As the size of the second phase particles increases, the particles impede dislocation movement and it becomes increasingly difficult for the particles to cut through the material. In other words, the strength of a material increases with increasing r.

Dislocations bowing around particle:

$$\tau = \frac{Gb}{L - 2r}$$

where τ is the material strength, G is the shear modulus, b is the magnitude of the Burgers vector, L is the distance between pinning points, and r is the second phase particle radius. This governing equation shows that for dislocation bowing the strength is inversely proportional to the second phase particle radius r. Dislocation bowing, also called Orowan strengthening, is more likely to occur when there are large particles present in the material.

These governing equations show that the precipitation hardening mechanism depends on the size of the precipitate particles. At small r, cutting will dominate, while at large r, bowing will dominate.

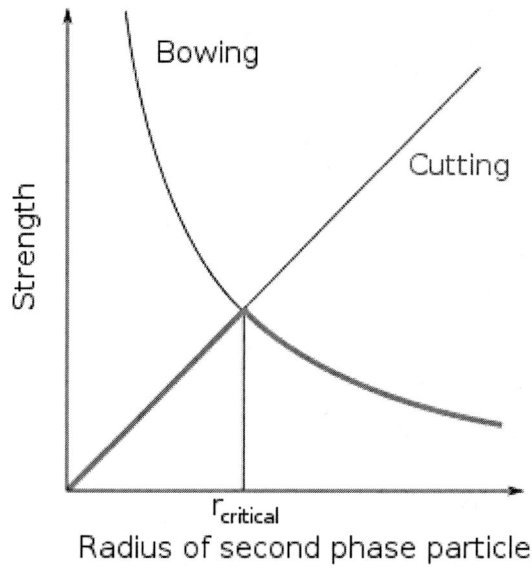

Figure 1 : Looking at the plot of both equations, it is clear that there is a critical radius at which max strengthening occurs. This critical radius is typically 5-30 nm.

Some Precipitation Hardening Materials

- 2000-series aluminum alloys (important examples: 2024 and 2019, also Y alloy and Hiduminium)
- 6000-series aluminum alloys (important example: 6061 for bicycle frames and aeronautical structures)
- 7000-series aluminum alloys (important examples: 7075 and 7475)
- 17-4PH stainless steel (UNS S17400)
- Maraging steel
- Inconel 718
- Alloy X-750
- René 41
- Waspaloy.

Quenching

In materials science, quenching is the rapid cooling of a workpiece to obtain certain material properties. It prevents low-temperature processes, such as phase transformations, from occurring by only providing a narrow window of time in which the reaction is both thermodynamically favorable and kinetically accessible. For instance, it can reduce crystallinity and thereby increase toughness of both alloys and plastics (produced through polymerization).

In metallurgy, it is most commonly used to harden steel by introducing martensite, in which case the steel must be rapidly cooled through its eutectoid point, the temperature at which austenite becomes unstable. In steel alloyed with metals such as nickel and manganese, the eutectoid temperature becomes much lower, but the kinetic barriers to phase transformation remain the same. This allows quenching to start at a lower temperature, making the process much easier. High speed steel also has added tungsten, which serves to raise kinetic barriers and give the illusion that the material has been cooled more rapidly than it really has. Even cooling such alloys slowly in air has most of the desired effects of quenching.

Extremely rapid cooling can prevent the formation of all crystal structure, resulting in amorphous metal or "metallic glass".

Quench Hardening

Quench hardening is a mechanical process in which steel and cast iron alloys are strengthened and hardened. These metals consist of ferrous metals and alloys. This is done by heating the material to a certain temperature, dependent upon material, and then rapidly cooling the material. This produces a harder material by either surface hardening or through-hardening varying on the rate at which the material is cooled. The material is then often tempered to reduce the brittleness that may increase from the quench hardening process. Items that may be quenched include gears, shafts, and wear blocks.

Process

Quenching metals is a progression; the first step is soaking the metal, i.e. heating it to the required temperature. Soaking can be done by air (air furnace), or a bath. The soaking time in air furnaces should be 1 to 2 minutes for each millimetre of cross-section. For a bath the time can range a little higher. The recommended time allotment in salt or lead baths is 0 to 6 minutes. Uneven heating or overheating should be avoided at all cost. Most materials are heated from anywhere to 815 to 900 °C (1,500 to 1,650 °F).

The next item on the progression list is the cooling of the part. Water is one of the most efficient quenching media where maximum hardness is acquired, but there is a small chance that it may cause distortion and tiny cracking. When hardness can be sacrificed, whale, cottonseed and mineral oils are used. These often tend to oxidize and form a sludge, which consequently lowers the efficiency. The quenching velocity (cooling rate) of oil is much less than water. Intermediate rates between water and oil can be obtained with water containing 10-30% Ucon, a substance with an inverse solubility which therefore deposits on the object to slow the rate of cooling.

To minimize distortion, long cylindrical workpieces are quenched vertically; flat workpieces are quenched on edge; and thick sections should enter the bath first. To prevent steam bubbles the bath is agitated.

Effect of Quench Hardening

Before the material is hardened, the microstructure of the material is a pearlite grain structure that is uniform and laminar. Pearlite is a mixture of ferrite and cementite formed when steel or cast iron are manufactured and cooled at a slow rate. After quench hardening, the microstructure of the material form into martensite as a fine, needle-like grain structure. Before using this technique it is essential to look up the rate constants for the quenching of the excited states of metal ions.

Equipment

There are three types of furnaces that are commonly used in quench hardening: salt bath furnace, continuous furnace, and box furnace. Each is used depending on what other processes or types of quench hardening are being done on the different materials.

Quenching Media

When quenching, there are numerous types of media. Some of the more common include: air, brine (salt water), oil and water. These media are used to increase the severity of the quench.

Tempering

Tempering is a heat treatment technique for metals, alloys and glass. In steels, tempering is done to "toughen" the metal by transforming brittle martensite or bainite into a combination of ferrite and cementite or sometimes Tempered martensite. Precipitation hardening alloys, like many grades of aluminum and superalloys, are tempered to precipitate intermetallic particles which strengthen the metal. Tempering is accomplished by a controlled reheating of the work piece to a temperature below its lower critical temperature.

The brittle martensite becomes tough and ductile after it is tempered. Carbon atoms were trapped in the austenite when it was rapidly cooled, typically by oil or water quenching, forming the martensite. The martensite becomes strong after being tempered because when reheated, the microstructure can rearrange and the carbon atoms can diffuse out of the distorted body-centred-tetragonal (BCT) structure. After the carbon diffuses, the result is nearly pure ferrite with body-centred structure.

In metallurgy and materials science, there is always a trade-off between strength and ductility. This delicate balance highlights many of the subtleties inherent to the tempering process. Precise control of time and temperature during the tempering process are critical to achieve a metal with well balanced mechanical properties.

Process characteristics

- Improves ductility and toughness
- Reduces cracking
- Improves machinability
- Increases impact resistance
- Improves malleability
- Decreases hardness.

Tempering in Steel

Typically steel is heat treated in a multi-step process. First it is heated to create a solid solution of iron and carbon in a process called *austenizing*. Austenizing is followed by quenching to produce a martensitic microstructure. The steel is then tempered by heating between the ranges of 150–260 °C (302–500 °F) and 370–650 °C (698–1202 °F). Tempering in the range of 260–370 °C (500–698 °F) is sometimes avoided to reduce temper brittling. The steel is held at that temperature until the carbon trapped in the martensite diffuses to produce a chemical composition with the potential to create either bainite or pearlite (a crystal structure formed from a mixture of ferrite and cementite). When producing a truly bainitic or pearlitic steel the steel must be once again taken up to the austenite region (austenizing) and cooled slowly to a controlled temperature before being fully quenched to a low temperature. In bainitic steels, upper bainite or lower bainite may form depending on the duration and temperature of the tempering process. It is thermodynamically impossible that the martensite will be totally converted during tempering, so a mixture of martensite, bainite, ferrite and cementite is often formed.

Tempering in Precipitation Hardened Alloys

Before a precipitation hardened alloy can be tempered, it must be placed in solution. During this process, the alloy is heated to dissolve and uniformly distribute alloying elements. The alloy is then quenched at a rate of cooling high enough to prevent the alloying elements from falling out of solution. The alloy is then tempered, by heating at temperatures lower than the dissolving temperature.

During tempering, the alloying elements will diffuse through the alloy and react to form intermetallic compounds. The intermetallic compounds are not soluble in the alloy, and will precipitate, forming small particles. These particles strengthen the metal by impeding the movement of dislocations through the crystal structure of the alloy. Careful manipulation of tempering time and temperature allows the size and amount of precipitates to be controlled, thus tailoring the mechanical properties of the alloy.

Tempering in aluminum is also referred to as "aging". Artificially aged alloys are tempered at elevated temperature, while naturally aging alloys may be tempered at room temperature.

Alloy systems with a large number of alloying elements, like some superalloys, may be subjected to several tempering operations. During each operation a different precipitate is formed, resulting in a large number of different precipitates that are difficult to drive back into solution. This phenomenon contributes to the high temperature strength of precipitation hardened superalloys.

Tempering in Blacksmithing

The temperatures used in tempering are often too low to be gauged by the colour of the workpiece. In this case, the blacksmith will heat the work piece for a known amount of time. Doing this ensures a certain degree of consistency in the tempering process from work piece to work piece. The cumulative effects of time and temperature can also be gauged by monitoring the colour of the oxide film formed while tempering a well-polished blade.

Tempered Glass

Tempering, or toughening, of glass is a process in which glass is first heated above its annealing temperature (about 720°C), and then rapidly cooled by jets of cool air, thus hardening the surface of the glass and leaving the centre soft for a period of time. This creates internal stresses in the glass. If the glass is broken, it will break into many small pieces instead of simply cracking, making it far safer. Another advantage is that tempered glass can be up to four times stronger than regular glass. However, tempered glass cannot be cut to shape or drilled once it has been toughened.

Tempering Colours

Steel that has been freshly ground or polished will form oxide layers when heated. At a very specific temperature, the iron oxide will form a layer with a very specfic thickness, causing thin-film interference. This causes colours to appear on the surface of the steel. As temperature is increased, the iron oxide layer grows in thickness, changing the colour. These colours, called tempering colours, have been used for centuries to gauge the temperature of the metal. At around 350°F (176°C) the steel will start to take on a very light, yellowish hue. At 400°F (204°C), the steel will become a noticable light-straw colour, and at 440°F (226°C), the colour will become dark-straw. At 500°F (260°C), steel will turn brown, while at 540°F (282°C) it will turn purple. At 590°F (310°C) the steel turns a very deep blue, but at 640°F (337°C) it becomes a rather light blue.

The tempering colours can be used to judge the final properties of the tempered steel. Very hard tool steel is often tempered in the light to dark straw range, whereas spring steel is often tempered to the blue. However, the final hardness of the tempered steel will vary, depending on the composition of the steel. The oxide film will also increase in thickness over time. Therefore, steel that has been held at 400°F for a very long time may turn brown or purple, even though the temperature never exceeded that needed to produce a light straw colour. Other factors affecting the final outcome are oil films on the surface and the type of heat source used.

Selective Hardening

Many heat treating methods have been developed to alter the

properties of only a portion of an object. These tend to consist of either cooling different areas of an alloy at different rates, by quickly heating in a localized area and then quenching, or by thermochemical diffusion.

Differential Hardening

Differential hardening is a method used in forging swords and knives to increase the hardness of the edge without making the whole blade brittle. To achieve this, the edge is cooled more rapidly than the spine by adding a heat insulator to the spine before quenching. Clay or another material is used for insulation. It can also be achieved by carefully pouring water (perhaps already heated) onto the edge of a blade as is the case with the manufacture of some kukri. This technique is mainly used in the Chinese jian and the katana, the traditional Japanese sword, and the khukuri, the traditional Nepalese knife. Most blades made with this technique have visible temper lines.

Another process, often referred to as differential hardening, but in reality differential *tempering* can also be obtained by quenching the object uniformly, then differentially tempering one part of it with a torch or some other directed heat source. The heated portion of the metal is softened by this process.

Flame Hardening

Flame hardening is used to harden only a portion of a metal. Unlike differential hardening, where the entire piece is heated and then cooled at different rates, in flame hardening, only a portion of the metal is heated before quenching. This is usually easier than differential hardening, but often produces an extremely brittle zone between the heated metal and the unheated metal, as cooling at the edge of this heat affected zone is extremely rapid.

Induction Hardening

Induction hardening is a form of heat treatment in which a metal part is heated by induction heating and then quenched. The quenched metal undergoes a martensitic transformation, increasing the hardness and brittleness of the part. Induction hardening is used to selectively harden areas of a part or assembly without affecting the properties of the part as a whole.

Process

Induction heating is a non contact heating process which utilises the principle of electromagnetic induction to produce heat inside the surface layer of a work-piece. By placing a conductive material into a strong alternating magnetic field electrical current can be made to flow in the steel thereby creating heat due to the I^2R losses in the material. In magnetic materials, further heat is generated below the curie point due to hysteresis losses. The current generated flows predominantly in the surface layer, the depth of this layer being dictated by the frequency of the alternating field, the surface power density, the permeability of the material, the heat time and the diameter of the bar or material thickness. By quenching this heated layer in water, oil or a polymer based quench the surface layer is altered to form a martensitic structure which is harder than the base metal.

Definition

A widely used process for the surface hardening of steel. The components are heated by means of an alternating magnetic field to a temperature within or above the transformation range followed by immediate quenching. The core of the component remains unaffected by the treatment and its physical properties are those of the bar from which it was machined, whilst the hardness of the case can be within the range 37/58 HRC. Carbon and alloy steels with an equivalent carbon content in the range 0.40/0.45% are most suitable for this process.

A source of high frequency electricity is used to drive a large alternating current through a coil. The passage of current through this coil generates a very intense and rapidly changing magnetic field in the space within the work coil. The workpiece to be heated is placed within this intense alternating magnetic field where eddy currents are generated within the workpiece and resistance leads to Joule heating of the metal.

This operation is most commonly used in steel alloys. Many mechanical parts, such as shafts, gears, and springs, are subjected to surface treatments, before the delivering, in order to improve wear behaviour. The effectiveness of these treatments depends both on surface materials properties modification and on the introduction of residual stress. Among these treatments, induction hardening is one of the most widely employed to improve component durability. It determines in the work-piece a tough core with tensile residual stresses and a hard surface layer with compressive stress, which have proved to be very effective in extending the component fatigue life and wear resistance. Induction surface hardened low alloyed medium carbon steels are widely used for critical automotive and machine applications which require high wear resistance. Wear resistance behaviour of induction hardened parts depends on hardening depth and the magnitude and distribution of residual compressive stress in the surface layer.

History

The basis of all induction heating systems was discovered in 1831 by Michael Faraday. Faraday proved that by winding two coils of wire around a common magnetic core it was possible to create a momentary emf in the second winding by switching the electric current in the first winding on and off. He further observed that if the current was kept constant, no emf was induced in the second winding and that this current flowed in opposite directions subject to whether the current was increasing or decreasing in the circuit.

Faraday concluded that an electric current can be produced by a changing magnetic field. As there was no physical connection between the primary and secondary windings, the emf in the secondary coil was said to be *induced* and so Faraday's law of induction was born. Once discovered, these principles were employed over the next century or so in the design of dynamos (electrical generators and electric motors, which are variants of the same thing) and in forms of electrical transformers. In these applications, any heat generated in either the electrical or magnetic circuits was felt to be undesirable. Engineers went to great lengths and used laminated cores and other methods to minimise the effects.

Early last century the principles were explored as a means to melt steel, and the motor generator was developed to provide the power required for the induction furnace. After general acceptance of the methodology for melting steel, engineers began to explore other possibilities for the utilisation of the process. It was already understood that the depth of current penetration in steel was a function of its magnetic permeability, resistivity and the frequency of the applied field. Engineers at Midvale Steel and The Ohio Crankshaft Company drew on this knowledge to develop the first surface hardening induction heating systems using motor generators.

The need for rapid easily automated systems led to massive advances in the understanding and utilisation of the induction hardening process and by the late 1950s many systems utilising motor generators and thermionic emission triode oscillators were in regular use in a vast array of industries. Modern day induction heating units utilise the latest in semiconductor technology and digital control systems to develop a range of powers from 1 kW to many megawatts.

Principal Methods

Single Shot Hardening

In single shot systems the component is held statically or rotated in the coil and the whole area to be treated is heated simultaneously for a pre-set time followed by either a flood quench or a drop quench system. Single shot is often used in cases where no other method will achieve the desired result for example for flat face hardening of hammers, edge hardening complex shaped tools or the production of small gears.

In the case of shaft hardening a further advantage of the single shot methodology is the production time compared with progressive traverse hardening methods. In addition the ability to use coils which can create longitudinal current flow in the component rather than diametric flow can be an advantage with certain complex geometry.

There are disadvantages with the single shot approach. The coil design can be an extremely complex and involved process. Often the use of ferrite or laminated loading materials is required to influence the magnetic field concentrations in given areas thereby to refine the heat pattern produced. Another drawback is that much more power is required due to the increased surface area being heated compared with a traverse approach.

Traverse Hardening

In traverse hardening systems the work piece is passed through the induction coil progressively and a following quench spray or ring is utilised. Traverse hardening is used extensively in the production of shaft type components such as axle shafts, excavator bucket pins, steering components, power tool shafts and drive shafts. The component is fed through a ring type inductor which normally features a single turn. The width of the turn is dictated by the traverse speed, the available power and frequency of the generator. This creates a moving band of heat which when quenched creates the hardened surface layer. The quench ring can be either integral a following arrangement or a combination of both subject to the requirements of the application. By varying speed and power it is possible to create a shaft which is hardened along its whole length or just in specific areas and also to harden shafts with steps in diameter or splines. It is normal when hardening round shafts to rotate the part during the process to ensure any variations due to concentricity of the coil and the component are removed. Traverse methods also feature in the production of edge components, such as paper knives, leather knives, lawnmower bottom blades, and hacksaw blades. These types of application normally utilise a hairpin coil or a transverse flux coil which sits over the edge of the component. The component is progressed through the coil and a following spray quench consisting of nozzles or drilled blocks.

Many methods are used to provide the progressive movement through the coil and both vertical and horizontal systems are used. These normally employ a digital encoder and programmable logic controller for the positional control, switching, monitoring, and setting. In all cases the speed of traverse needs to be closely controlled and consistent as variation in speed will have an effect on the depth of hardness and the hardness value achieved.

Equipment

Power Required

Power supplies for induction hardening vary in power from a few kilowatts to hundreds of kilowatts dependent of the size of the component to be heated and the production method employed i.e. single shot hardening, traverse hardening or submerged hardening.

In order to select the correct power supply it is first necessary to calculate the surface area of the component to be heated. Once this has been established then a variety of methods can be used to calculate the power density required, heat time and generator operating frequency. Traditionally this was done using a series of graphs, complex empirical calculations and experience. Modern techniques typically utilise finite element analysis and Computer-aided manufacturing techniques, however as with all such methods a thorough working knowledge of the induction heating process is still required.

For single shot applications the total area to be heated needs to be calculated. In the case of traverse hardening the circumference of the component is multiplied by the face width of the coil. Care must be exercised when selecting a coil face width that it is practical to construct the coil of the chosen width and that it will live at the power required for the application.

Frequency

Induction heating systems for hardening are available in a variety of different operating frequencies typically from 1 kHz to 400 kHz. Higher and lower frequencies are available but typically these will be used for specialist applications. The relationship between operating frequency and current penetration depth and therefore hardness depth is inversely proportional. i.e. the lower the frequency the deeper the case.

Table : Examples of frequencies for various case depths and material diameters

Case depth [mm]	*Bar diameter [mm]*	*Frequency [kHz]*
0.8 to 1.5	5 to 25	200 to 400
1.5 to 3.0	10 to 50	10 to 100
	>50	3 to 10
3.0 to 10.0	20 to 50	3 to 10
	50 to 100	1 to 3
	>100	1

The above table is purely illustrative, good results can be obtained outside these ranges by balancing power densities, frequency and other practical considerations including cost which may influence the final selection, heat time and coil width. As well as the power density and frequency, the time the material is heated for will influence the depth to which the heat will flow by conduction. The time in the coil can be influenced by the traverse speed and the coil width, however this will also have an effect on the overall power requirement or the equipment throughput.

It can be seen from the above table that the selection of the correct equipment for any application can be extremely complex as more than one combination of power, frequency and speed can be used for a given result. However in practice many selections are immediately obvious based on previous experience and practicality.

Case Hardening

Case hardening or surface hardening is the process of hardening the surface of a metal, often a low carbon steel, by infusing elements into the material's surface, forming a thin layer of a harder alloy. Case hardening is usually done after the part in question has been formed into its final shape, but can also be done to increase the hardening element content of bars to be used in a pattern welding or similar process. The term face hardening is also used to describe this technique, when discussing modern armour.

History

Early iron melting made use of bloomeries, which produced two layers of metal: one with a very low carbon content that is worked into wrought iron, and the rest a high carbon cast iron. Since the high carbon iron is *hot short*, meaning it fractures and crumbles when forged, it was not useful without more smelting. The wrought iron, with nearly no carbon in it, was very malleable and ductile, but not very hard.

Case hardening involves packing the low-carbon iron within a substance high in carbon, then heating this pack to encourage carbon migration into the surface of the iron. This forms a thin surface layer of higher carbon steel, with the carbon content gradually decreasing deeper from the surface. The resulting product combines much of the toughness of a low-carbon steel core, with the hardness and wear resistance of the outer high-carbon steel.

The traditional method of applying the carbon to the surface of the iron involved packing the iron in a mixture of ground bone and charcoal, or a combination of leather, hooves, salt and urine, all inside a well-sealed box. This carburizing package is then heated to a high temperature, but still under the melting point of the iron, and left at that temperature for a length of time. The longer the package is held at the high temperature, the deeper the carbon will diffuse into the surface. Different depths of hardening is desirable for different purposes: sharp tools need deep hardening to allow grinding and resharpening without exposing the soft core, while machine parts like gears might need only shallow hardening for increased wear resistance.

The resulting case hardened part may show distinct surface discoloration. The steel darkens significantly, and shows a mottled pattern of black, blue and purple, caused by the various compounds formed from impurities in the bone and charcoal. This oxide surface works similarly to bluing, providing a degree of corrosion resistance, as well as an attractive finish. *Case colouring* refers to this pattern and is commonly encountered as a decorative finish on replica historic firearms.

With modern steelworking techniques, it is possible to make homogeneous steels of low to high carbon content, removing much of the original motivation for case hardening. However, the heterogeneous nature of case hardened steel may still be desirable, as it can combine both extreme hardness and extreme toughness, something which is not readily matched by homogeneous alloys.

Chemistry

Carbon itself is solid at case-hardening temperatures and so is immobile. Transport to the surface of the steel was as gaseous carbon monoxide, generated by the breakdown of the carburising compound and the oxygen packed into the sealed box. This takes place with pure carbon, but unworkably slowly. Although oxygen is required for this process it's re-circulated through the CO cycle and so can be carried out inside a sealed box. The sealing is necessary to stop the CO either leaking out, or being oxidised to CO_2 by excess outside air.

Adding an easily decomposed carbonate "energiser" such as barium carbonate breaks down to BaO + CO_2 and this encourages the reaction

C (from the donor) + CO_2 <—> 2 CO

increasing the overall abundance of CO and the activity of the carburising compound. It's 'common knowledge' that case-hardening was done with bone, but this is misleading. Although bone was used, the main carbon donor was hoof and horn. Bone contains some carbonates, but is mainly calcium phosphate (as hydroxylapatite). This doesn't have the beneficial effect on encouraging CO production and it can also supply phosphorus as an impurity into the steel alloy.

Modern use

Both carbon and alloy steels are suitable for case-hardening; typically mild steels are used, with low carbon content, usually less than 0.3%. These mild steels are not normally hardenable due to the low quantity of carbon, so the surface of the steel is chemically altered to increase the hardenability. Case hardened steel is formed by diffusing carbon (carburization), nitrogen (nitriding) and/or boron (boriding) into the outer layer of the steel at high temperature, and then heat treating the surface layer to the desired hardness.

The term case hardening is derived from the practicalities of the carburization process itself, which is essentially the same as the ancient process. The steel work piece is placed inside a case packed tight with a carbon-based case hardening compound. This is collectively known as a carburizing pack. The pack is put inside a hot furnace for a variable length of time. Time and temperature determines how deep into the surface the hardening extends. However, the depth of hardening is ultimately limited by the inability of carbon to diffuse deeply into solid steel, and a typical depth of surface hardening with this method is up to 1.5 mm. Other techniques are also used in modern carburizing, such as heating in a carbon-rich atmosphere. Small items may be case hardened by repeated heating with a torch and quenching in a carbon rich medium, such as the commercial products *Kasenit/Casenite* or "Cherry Red". Older formulations of these compounds contain potentially toxic cyanide compounds, such as ferrocyanide compounds, while the more recent types such as Cherry Red do not.

Flame and Induction Hardening

Flame or induction hardening are processes in which the surface of the steel is heated to high temperatures (by direct application of a flame, or by induction heating) then cooled rapidly, generally using water; this creates a "case" of martensite on the surface. A carbon content of 0.4–0.6 wt% C is needed for this type of hardening.

Typical uses are for the shackle of a lock, where the outer layer is hardened to be file resistant, and mechanical gears, where hard gear mesh surfaces are needed to maintain a long service life while toughness is required to maintain durability and resistance to catastrophic failure.

Bibliography

Albritton, C. C. : *The Fabric of Geology: Reading*, Mass., Addison-Wesley Publishing Co., 1963.

Armistead, George: *Safety in Petroleum Refining and Related Industries*, New York: John G. Simmons & Co., 1950.

Arora, G D : *Liquid Crystals and Polymers*, Sarup and Sons, Delhi, 2005.

Arun Kumar Biswas: *Minerals and Metals in Pre-Modern India*, D K Printworld, Delhi, 2001.

Balakrishan, A. : *Biofuel Crops and Renewable Energy*, Sublime Publications, Delhi, 2011.

Barker, Daniel S.: *Igneous Rocks*, Prentice-Hall, Delhi, 1983.

Barnes, Thomas G. : *Origin and Destiny of the Earth s Magnetic Field*, Creation Research Society, London, 1983.

Belknap, D.F.: *Sea-level Rise on the Maine Islands: Proceedings*, Sixth Island Institute Conference, Hurricane Island, Maine, Island Institute, Rockland, ME, 1988.

Berlin, G.L.: *Earthquakes and the Urban Environment*, CRC Press, Boca Raton, FL, 1980.

Bhakar S.R. : *Ground Water Hydrology : Theory and Practice*, Agrotech, Delhi, 2009.

Biggar, John: *The Andes: A Guide for Climbers*, Andes Publishing, Scotland, 2005.

Binay Kumar and R.P. Tandon: *Advances in Technologically Important Crystals*, Macmillan Publishers India, Delhi, 2007.

Biyani, A.K. : *Dimensions of Himalayan Geology*, Satish Serial Pub, Delhi, 2007.

Bogli, Alfred: *Karst Hydrology and Physical Speleology*, Springer-Verlag, New York, 1980.

Bromley, Simon: *American Hegemony and World Oil: The Industry, the State System, and the World Economy*. University Park, Pa., 1991.

Chowdhary, L.R. : *Petroleum Geology of the Cambay Basin, Gujarat, India*, Indian Petroleum, Delhi, 2004.

Cotton, C.A.: *Volcanoes as Landscape Forms*, Hafner Publishing. New York, 1999.

DeLaisi, Francis: *Oil, Its Influence on Politics*. Labor Publishing,1923.

Doyle, F. L.: *Karst Hydrogeology,* The University of Alabama in Huntsville Press, Huntsville, Alabama, 1977.

Dreybrodt, Wolfgang: *Processes in Karst Systems Physics, Chemistry and Geology,* Springer-Verlag, New York, 1988.

Dunai, T.J.: *Cosmogenic Nucleides,* Cambridge University Press, U.K., 2010.

Eaton, Rev. S.J.M.: *Petroleum, a History of the Oil Region*, J.P. Skelly & Co., Philadelphia, 1866.

Edward W.: *United States Oil Policy and Diplomacy: A Twentieth-Century Overview*. West-port, Conn.: Greenwood Press, 1983.

Edwards, Denis : *Earth Revealing Earth Healing: Ecology and Christian Theology,* Collegeville: Liturgical Press, 2001.

Edwards, Edward : *Pattern and Design with Dynamic Symmetry,* NY: Dover 1967.

Fischer, Louis: *Oil Imperialism: The International Struggle for Petroleum.* New York: International Publishers, 1926.

Folk, R.L.: *The Petrology of Sedimentary Rocks*, Hemphill Publishing Co., Austin, 2001.

Fritz Helmet: *Diffraction in Crystals*, Ivy Publishing House, Delhi, 2011.

Garverick, Linda: *Corrosion in the Petrochemical Industry*. Materials Park, Ohio: ASM International, 2000.

Ghosh, N.C. and K.D. Sharma: *Groundwater Modelling and Management*, Capital Pub, Delhi, 2006.

Glenn T. Trewartha : *Elements of Geography*, New York, McGraw Hill, 1941.

Gontz, A.M.: *Evolution of Seabed Pockmarks in Penobscot Bay, Maine*, University of Maine, Orono, 2011.

Gubbins, David: *Seismology and Plate Tectonics*, Cambridge University Press, North America, 1990.

Günter Simader: *Fuel Cells and Their Applications*. New York: VCH, 1996.

Gurjar, Ram Kumar : *Geography of Water Resources*, Rawat, Delhi, 2008.

Gurugnanam, B. : *Essentials of Hydrogeology*, New India Pub, Delhi, 2009.

Hall, Anthony: *Igneous Petrology,* Longman, New York, 1996.

Hayward, J.A.: *Hydrology and Stream Sediment from Torlesse Stream Catchment*, Oxford University, UK, 1980.

Holmgren, D. A., Moody, J. D. : *The Structural Setting for Giant oil and Gas Fields*, Tokyo, 1975.

Inkpen, Robert: *Science, Philosophy and Physical Geography*, Routledge, London.

Jakucs, Laszlo: *Morphogenetics of Karst Regions: Variants of Karst Evolution,* Halstead Press, John Wiley and Sons, New York, 1977.

John C. Grover.: *Volcanic Eruptions and Great Earthquakes*, Copyright Publishing Company, Australia, 1998.

Kumar Yadav, Surendra : *Soil Ecology : Remote Sensing and Geographic Information System in Management,* APH, Delhi, 2007.

Lovesun, V.J. ; P.K. Sen and Amalendu Sinha: *Exploration, Exploitation, Enrichment and Environment of Coastal Placer Minerals*, Macmillan Publishers India, Delhi, 2007.

Mahajan Gautam : *Ground Water : Surveys and Investigation*, APH, Delhi, 2009.

Majumdar P.K. Basu: *Law of Mines and Minerals*, Universal Law, Delhi, 2011.

Malhotra, Raman : *Mines and Minerals Laws*, Global India, Delhi, 2011.

Mallik, T K : *Marine Geology : A Scenario Around Indian Coasts*, New Academic Pub, Delhi, 2008.

Mandelbrot, Benoit : *The Fractal Geometry of Nature,* San Francisco: Freeman, 1982.

Manthorp, P.A.: *Estuarine Circulation and Sediment Transport in the Saco River estuary*, Boston University, Boston, MA, 1995.

Mary Lambert: *Crystal Energy*, Cico Books, Delhi, 2010.

McLaurin, John J.: *Sketches in Crude-oil*, J. Horace McFarland Company, Harrisburg, 1896.

Michael Mugerwa: *Fuel Cell Systems*. New York: Plenum Press, 1993.

Middlemost, Eric A. K.: *Magmas and Magmatic Rocks: An Introduction to Igneous Petrology*, Longman, New York, 1985.

Miller, J.R.: *Process Geomorphology*, Wm.C. Brown Publishers, Dubuque, Iowa, 1995.

Poonam Johri: *Minerals and Water*, Sonali, Delhi, 2005.

Postnikov, M. : *Lectures in Geometry Semester I Analytic Geometry*, Moscow: MIR publishers, 1982.

Pramod O. Alexander: *A Handbook of Minerals, Crystals, Rocks and Ores*, New India Pub Agency, Delhi, 2009.

Ramasamy, S. : *Remote Sensing in Water Resources,* Rawat, 2005.

Rathi, Rakesh : *Petroleum Refining Processes*, SBS Pub, Delhi, 2007.

Raymond M. : *Laser Remote Sensing: Fundamentals and Applications*, Malabar, Fla., Krieger Pub. 1992.

Reiter, L.: *Earthquake Hazard Analysis: Issues and Insights*, Columbia University Press, New York, 1990.

Richards, John A. : *Remote Sensing Digital Image Analysis: An Introduction*, Berlin, New York, 1993.

Samuel Delvin: *Vitamins, Minerals and Hormones*, Ivy Pub, Delhi, 2003.

Santhi, R and K M Sellamuthu: *Fundamentals of Forest Soils*, Satish Serial Pub, Delhi, 2008.

Sawant, P. T. : *Engineering and General Geology*, New India Publishing Agency, Delhi, 2011.

Selby, Michael John: *Earth s Changing Surface: an Introduction to Geomorphology*, Clarendon Press, Oxford, 1985.

Shearer, Peter M.: *Introduction to Seismology*, Cambridge University Press, North America, 1998.

Sidney Russell Cooke: *The Oil Trusts and Anglo-American Relations*. Macmillan, 1924

Speight, James G.: *The Chemistry and Terminology of Petroleum*. New York: Marcel Dekker, 1980.

Steven, A. : *Grand Canyon: Monument to Catastrophe,* Institute for Creation Research, London,1994.

Summerfield, M.A.: *Global Geomorphology*, Pearson Education Ltd, Delhi, 1991.

Talukdar, A. K. : *Catalysis in Petroleum and Petrochemical Industries*, Alpha Science International, Delhi, 2005.

Tiwari, S.K. : *Ore Geology, Economic Minerals And Mineral Economics*, Atlantic Publishers, Delhi, 2010.

Tracy, Robert J.; Owens, Brent: *Petrology: Igneous, Sedimentary, and Metamorphic,* New York: W. H. Freeman, 2005.

Vijayalakshmi, G S : *Fuels and Biofuels*, Agrobios, Delhi, 2007.

Wahab, M. A. : *Essentials of Crystallography*, Narosa, Delhi, 2009.

Wahrhaftig, Clyde: *A Streetcar to Subduction: and Other Plate Tectonic Trips by Public Transportation in San Francisco*, American Geophysical Union, Washington, D.C., 1984.

Wainwright, John; Mulligan, M.: *Environmental Modelling: Finding Simplicity in Complexity*, John Wiley and Sons Ltd, London, 2003.

White, William B.: *Geomorphology and Hydrology of Karst Terrains,* Oxford University Press, New York, 1988.

Whitmore, G : *Chemical Crystallography And Lipid Crystals*, Sarup Book, Delhi, 2009.

Williams, Paul: *Karst Geomorphology and Hydrology,* Unwin Hyman, Winchester, Massachusetts, 1989.

Young, George J.: *Fuel Cells,* New York: Reinhold Publishing Corp., 1961.

Index

□□□